STUDY GUIDE

Donald G. Newnan

Ed Wheeler

for

ENGINEERING ECONOMIC ANALYSIS

NINTH EDITION

Donald G. Newnan

Ted G. Eschenbach

Jerome P. Lavelle

New York Oxford

OXFORD UNIVERSITY PRESS

2004

Oxford University Press

Oxford New York
Auckland Bangkok Buenos Aires Cape Town Chennai
Dar es Salaam Delhi Hong Kong Istanbul Karachi Kolkata
Kuala Lumpur Madrid Melbourne Mexico City Mumbai
Nairobi São Paulo Shanghai Taipei Tokyo Toronto

Published by Oxford University Press, Inc.
198 Madison Avenue, New York, New York, 10016
http://www.oup-usa.org

ISBN 0–19–517149–7

Printing number: 9 8 7 6 5 4 3 2

Printed in the United States of America
on acid-free paper

Contents

Foreword iv
Formulas v

Course Summary 1

1 Making Economic Decisions 37

2 Engineering Costs and Cost Estimating 45

3 Interest and Equivalence 49

4 More Interest Formulas 53

5 Present Worth Analysis 71

6 Annual Cash Flow Analysis 95

7 Rate of Return Analysis 121

7A Difficulties Solving for an Interest Rate 131

8 Incremental Analysis 133

9 Other Analysis Techniques 137

10 Uncertainty in Future Events 161

11 Depreciation 169

12 Income Taxes 179

13 Replacement Analysis 199

14 Inflation and Price Change 209

15 Selection of a Minimum Attractive Rate of Return 225

16 Economic Analysis in the Public Sector 229

17 Rationing Capital Among Competing Projects 241

18 Accounting and Engineering Economy 245

Compound Interest Tables 251

Foreword

Over the years I have determined that students often do not have study materials that are truly useful to them. They usually have the course textbook and the lecture notes that they have managed to take or someone has take for them. Often these lecture notes leave much to be desired in terms of clear, organized, useful information. Occasionally a student will have copies of old exams given by their professor. These exams can be treasure troves of information for the student. For the past several years I have made my old quizzes and exams available to my students and have been told that they are the most useful of all the resources available for studying for an exam.

This study guide is the product of the revision of a previous guide and the addition of new material. There are now problems covering every chapter in the companion text, *Engineering Economic Analysis*. The majority of them are actual exam problems used in an introductory course in engineering economy. In general, the exam problems are just as they appeared on the actual course examination. Problems were carefully selected for publication so that the fundamentals of engineering economy found in your textbook are covered.

This book begins with a summary of basic engineering economy principles. The main section of the book consists of over 350 problems with solutions. The problems offer a wide range of complexity and different solution formats. The last section contains a set of compound interest tables typically used in the solution of engineering economy problems.

Thank you to the following engineering economy professors from across the country who have contributed problems to the earlier version of this guide: Hal Ball, B.G. Barr, Charles Buford (deceased), Gary Crossman, Anthony DeFruscio, R. James Diegel, John Eastman, Bryan Jenkins, Jack Lohmann, Hugo Patino, Joseph Pignatiello, Jr., Paul Schonfeld, Barry Shiller, Ralph Smith, and Thomas Ward.

I would like to thank Dr. Ted G. Eschenbach for recommending me to the publisher of this guide. And I would like to thank Danielle Christensen, Jenny Boully, and Karen Shapiro —all of Oxford University Press— for their encouragement and help in preparing this book. A special thanks goes to Michelle Drumwright who retyped the previous edition when electronic copies of it could not be found. And the biggest thanks of all go to my wife Ellen and daughter Abigail for their patience and understanding while I worked on this project.

If you find any errors in the problems or solutions, please inform the editor in care of the Department of Engineering at the University of Tennessee at Martin or by e-mail at ewheeler@utm.edu. Also if you would like to submit a problem or problems for a future edition please use the same addresses. I hope you find this material facilitates a better understanding of engineering economy and that it helps you to succeed in your course!

Ed Wheeler
Coeditor

Formulas

Compound Amount: To find F, given P

$(F/P, i, n)$ $$F = P(1+i)^n$$

Present Worth: To find P, given F

$(P/F, i, n)$ $$P = F(1+i)^{-n}$$

Series Compound Amount: To find F, given A

$(F/A, i, n)$ $$F = A\left[\frac{(1+i)^n - 1}{i}\right]$$

Sinking Fund: To find A, given F

$(A/F, i, n)$ $$A = F\left[\frac{i}{(1+i)^n - 1}\right]$$

Capital Recovery: To find A, given P

$(A/P, i, n)$ $$A = P\left[\frac{i(1+i)^n}{(1+i)^n - 1}\right]$$

Series Present Worth: To find P, given A

$(P/A, i, n)$ $$P = A\left[\frac{(1+i)^n - 1}{i(1+i)^n}\right]$$

Arithmetic Gradient Uniform Series: To find A, given G

$(A/G, i, n)$ $$A = G\left[\frac{(1+i)^n - in - 1}{i(1+i)^n - i}\right] \quad \text{or} \quad A = G\left[\frac{1}{i} - \frac{n}{(1+i)^n - 1}\right]$$

Arithmetic Gradient Present Worth: To find P, given G

$(P/G, i, n)$ $$P = G\left[\frac{(1+i)^n - in - 1}{i^2(1+i)^n}\right]$$

Geometric Gradient: To find P, given A_1, g

$(P/G, g, i, n)$

$$P = A_1\left[n(1+i)^{-1}\right]$$

when $i = g$

$$P = A_1\left[\frac{1-(1+g)^n(1+i)^{-n}}{i-g}\right]$$

when $i \neq g$

Continuous Compounding at Nominal Rate r

Single Payment: $$F = P\left[e^{rn}\right] \qquad P = F\left[e^{-rn}\right]$$

Uniform Series: $$A = F\left[\frac{e^r - 1}{e^{rn} - 1}\right] \qquad A = P\left[\frac{e^{rn}(e^r - 1)}{e^{rn} - 1}\right]$$

$$F = A\left[\frac{e^{rn} - 1}{e^{r} - 1}\right] \qquad P = A\left[\frac{e^{rn} - 1}{e^{rn}(e^{r} - 1)}\right]$$

Compound Interest —*Normally the interest period is one year, but it could be something else.

- ***i*** = Interest rate per interest period*.
- ***n*** = Number of interest periods.
- ***P*** = A present sum of money.
- ***F*** = A future sum of money.
- ***A*** = An end-of-period cash receipt or disbursement in a uniform series continuing for *n* periods.
- ***G*** = Uniform period-by-period increase or decrease in cash receipts or disbursements.
- ***g*** = Uniform rate of cash flow increase or decrease from period to period; the geometric gradient.
- ***r*** = Nominal interest rate per interest period*.
- ***m*** = Number of compounding subperiods per periods*.

Effective Interest Rates

For non-continuous compounding: i_{eff} or $i_a = \left(1 + \frac{r}{m}\right)^{m} - 1$

where r = nominal interest rate per year
m = number of compounding periods in a year

OR

i_{eff} or $i_a = (1 + i)^{m} - 1$

where i = effective interest rate per period
m = number of compounding periods in a year

For continuous compounding: i_{eff} or $i_a = (e^{r}) - 1$

where r = nominal interest rate per year

Values of Interest Factors When n Equals Infinity

Single Payment:

$(F/P, i, \infty) = \infty$
$(P/F, i, \infty) = 0$

Uniform Payment Series:

$(A/F, i, \infty) = 0$
$(A/P, i, \infty) = i$
$(F/A, i, \infty) = \infty$
$(P/A, i, \infty) = 1/i$

Arithmetic Gradient Series:

$(A/G, i, \infty) = 1/i$
$(P/G, i, \infty) = 1/i^2$

COURSE SUMMARY

This chapter is a brief review of engineering economic analysis/engineering economy. The goal is to give you a better grasp of the major topics in a typical first course. Hopefully this overview will help you put the course lectures and your reading of the textbook in better perspective. There are 26 example problems scattered throughout the engineering economics review. These examples are an integral part of the review and should be worked to completion as you come to them.

CASH FLOW

The field of engineering economics uses mathematical and economic techniques to systematically analyze situations that pose alternative courses of action.

The initial step in engineering economics problems is to resolve a situation, or each alternative course in a given situation, into its favorable and unfavorable consequences or factors. These are then measured in some common unit -- usually money. Those factors that cannot readily be reduced to money are called intangible, or irreducible, factors. Intangible or irreducible factors are not included in any monetary analysis but are considered in conjunction with such an analysis when making the final decision on proposed courses of action.

A cash flow table shows the "money consequences" of a situation and its timing. For example, a simple problem might be to list the year-by-year consequences of purchasing and owning a used car:

Year	*Cash Flow*	
Beginning of first Year 0	-4500	Car purchased "now" for $4500 cash. The minus sign indicates a disbursement.
End of Year 1	-350	Maintenance costs are $350 per year
End of Year 2	-350	
End of Year 3	-350	
End of Year 4	-350	
	+2000	The car is sold at the end of the 4th year for $2000. The plus sign represents a receipt of money.

This same cash flow may be represented graphically:

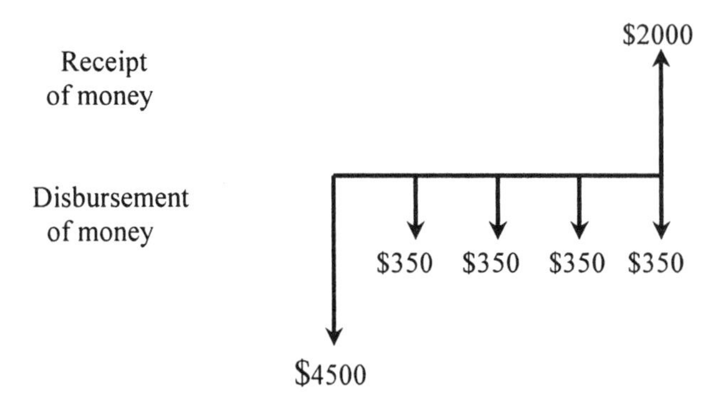

The upward arrow represents a receipt of money, and the downward arrows represent disbursements.

The x-axis represents the passage of time.

EXAMPLE 1

In January 1999, a firm purchases a used typewriter for $500. Repairs cost nothing in 1999 or 2000. Repairs are $85 in 2001, $130 in 2002, and $140 in 2003. The machine is sold in 2003 for $300. Compute the cash flow table.

Solution

Unless otherwise stated in problems, the customary assumption is a beginning-of-year purchase, followed by end-of-year receipts or disbursements, and an end-of-year resale or salvage value. Thus the typewriter repairs and the typewriter sale are assumed to occur at the end of the year. Letting a minus sign represent a disbursement of money, and a plus sign a receipt of money, we are able to set up this cash flow table:

Year	*Cash Flow*
Beginning of 1999	-$500
End of 1999	0
End of 2000	0
End of 2001	-85
End of 2002	-130
End of 2003	+160

Notice that at the end of 1994, the cash flow table shows +160. This is the net of -140 and +300.

If we define Year 0 as the beginning of 1990, the cash flow table becomes:

Year	*Cash Flow*
0	-$500
1	0
2	0
3	-85
4	-130
5	+160

From this cash flow table, the definitions of Year 0 and Year 1 become clear. Year 0 is defined as the beginning of Year 1. Year 1 is the end of Year 1. Year 2 is the *end* of Year 2, and so forth.

TIME VALUE OF MONEY

When the money consequences of an alternative occur in a short period of time -- say, less than one year -- we might simply add up the various sums of money and obtain the net result. But we cannot treat money this same way over longer periods of time. This is because money today is not the same as money at some future time.

Consider this question: Which would you prefer, $100 today or the assurance of receiving $100 a year from now? Clearly, you would prefer the $100 today. If you had the money today, rather than a year from now, you could use it for the year. And if you had no use for it, you could lend it to someone who would pay interest for the privilege of using your money for the year.

EQUIVALENCE

In the preceding section we saw that money at different points in time(for example, $100 today or $100 one year hence) may be equal in the sense that they both are $100, but $100 a year hence is not an acceptable substitute for $100 today. When we have acceptable substitutes, we say they are *equivalent* to each other. Thus at 8% interest, $108 a year hence is equivalent to $100 today.

EXAMPLE 2

At a 10% per year interest rate, $500 now is *equivalent* to how much three years hence?

Solution

$500 now will increase by 10% in each of the three years.

Now	=		$500.00
End of 1st year	= 500 + 10%(500)	=	550.00
End of 2nd year	= 550 + 10%(550)	=	605.00
End of 3rd year	= 605 + 10%(605)	=	665.50

Thus $500 now is *equivalent* to $665.50 at the end of three years.

Equivalence is an essential factor in engineering economic analysis. Suppose we wish to select the better of two alternatives. First, we must compute their cash flows. An example would be:

	Alternative	
Year	*A*	*B*
0	-$2000	-$2800
1	+800	+1100
2	+800	+1100
3	+800	+1100

The larger investment in Alternative *B* results in larger subsequent benefits, but we have no direct way of knowing if Alternative *B* is better than Alternative *A*. Therefore we do not know which alternative should be selected. To make a decision we must resolve the alternatives into *equivalent* sums so they may be compared accurately and a decision made.

COMPOUND INTEREST FACTORS

To facilitate equivalence computations a series of compound interest factors will be derived and their use illustrated.

Symbols

i = Interest rate per interest period. In equations the interest rate is stated as a decimal(that is, 8% interest is 0.08).

n = Number of interest periods.

P = A present sum of money.

F = A future sum of money. The future sum F is an amount, n interest periods from the present, that is equivalent to P with interest rate i.

A = An end-of-period cash receipt or disbursement in a uniform series continuing for n periods, the entire series equivalent to P or F at interest rate i.

G = Uniform period-by-period increase in cash flows; the arithmetic gradient.

g = Uniform *rate* of period-by-period increase in cash flows; the geometric gradient.

Functional Notation

	To Find	Given	Functional Notation
Single Payment			
Compound Amount Factor	F	P	$(F/P, i, n)$
Present Worth Factor	P	F	$(P/F, i, n)$
Uniform Payment Series			
Sinking Fund Factor	A	F	$(A/F, i, n)$
Capital Recovery Factor	A	P	$(A/P, i, n)$
Compound Amount Factor	F	A	$(F/A ,i ,n)$
Present Worth Factor	P	A	$(P/A , i, n)$
Arithmetic Gradient			
Gradient Uniform Series	A	G	$(A/G, i, n)$
Gradient Present Worth	P	G	$(P/G, i, n)$

From the table above we can see that the functional notation scheme is based on writing(To Find / Given, i, n). Thus, if we wished to find the future sum F, given a uniform series of receipts A, the proper compound interest factor to use would be$(F/A, i, n)$.

Single Payment Formulas

Suppose a present sum of money P is invested for one year at interest rate i. At the end of the year, we receive back our initial investment P together with interest equal to Pi or a total amount P + Pi. Factoring P, the sum at the end of one year is P(1 + i). If we agree to let our investment remain for subsequent years, the progression is as follows:

	Amount at Beginning of Period	+	Interest for the Period	=	Amount at End of the Period
1st year	P	+	Pi	=	$P(1 + i)$
2nd year	$P(1 + i)$	+	$Pi\,P(1 + i)$	=	$P(1 + i)^2$
3rd year	$P(1 + i)^2$	+	$Pi\,P(1 + i)^2$	=	$P(1 + i)^3$
nth year	$P(1 + i)^{n-1}$	+	$Pi\,P(1 + i)^{n-1}$	=	$P(1 + i)^n$

The present sum P increases in n periods to $P(1 + i)^n$. This gives us a relationship between a present sum P and its equivalent future sum F:

Future Sum =(Present Sum)$(1 + i)^n$

$$F = P(1 + i)^n$$

This is the Single Payment Compound Amount formula. In functional notation it is written:

$$F = P(F/P, i, n)$$

The relationship may be rewritten as:

Present Sum =(Future Sum)$(1 + i)^{-n}$

$$P = F(1 + i)^{-n}$$

This is the Single Payment Present Worth formula. It is written:

$$P = F(P/F, i, n)$$

EXAMPLE 3

At a 10% per year interest rate, $500 now is equivalent to how much three years hence?

Solution

This problem was solved in Example 2. Now it can be solved using a single payment formula.

$P = \$500$ $F = \text{unknown}$
$n = 3 \text{ years}$ $i = 10\%$

$$F = P(1 + i)^n = 500(1 + 0.10)^3 = \$665.50$$

This problem may also be solved using the Compound Interest Tables.

$$F = P(F/P, i, n) = 500(F/P, 10\%, 3)$$

From the 10% Compound Interest Table, read$(F/P, 10\%, 3) = 1.331$.

$$F = 500(F/P, 10\%, 3) = 500(1.331) = \$665.50$$

EXAMPLE 4

To raise money for a new business, a man asks you to loan him some money. He offers to pay you $3000 at the end of four years. How much should you give him now if you want 12% interest per year on your money?

Solution

$F = \$3000$ $P = \text{unknown}$
$n = 4 \text{ years}$ $i = 12\ \%$

$$P = F(1 + i)^{-n} = 3000(1 + 0.12)^{-4} = \$1906.55$$

Alternate computation using Compound Interest Tables:

$$\begin{aligned} P = F(P/F, i, n) &= 3000(P/F, 12\%, 4) \\ &= 3000(0.6355) \\ &= \$1906.50 \end{aligned}$$

Note that the solution based on the Compound Interest Table is slightly different from the exact solution using a hand calculator. In economic analysis, the Compound Interest Tables are always considered to be sufficiently accurate.

Uniform Payment Series Formulas

A uniform series is identical to n single payments, where each single payment is the same and there is one payment at the end of each period. Thus, the present worth of a uniform series is derived algebraically by summing n single-payments. The derivation of the equation for the present worth of a uniform series is shown below.

$$P = A\,[\quad 1/(1+i)^1 + 1/(1+i)^2 + \ldots + 1/(1+i)^{n-1} + 1/(1+i)^n]$$
$$(1+i)\,P = A\,[1/(1+i)^0 + 1/(1+i)^1 + 1/(1+i)^2 + \ldots + 1/(1+i)^{n-1}]$$
$$(1+i)P - P = A\,[1/(1+i)^0 \quad - 1/(1+i)^n]$$
$$iP = A\,[1 - 1/(1+i)^n]$$
$$= A\,[(1+i)^n - 1]/(1+i)^n]$$
$$P = A\,[(1+i)^n - 1]\,/\,[i(1+i)^n] \quad \text{Uniform Series Present worth formula}$$

Solving this equation for A:

$$A = P\,[i(1+i)^n]\,/\,[(1+i)^n - 1] \quad \text{Uniform Series Capital Recovery formula}$$

Since $F = P(1+i)^n$, we can multiply both sides of the P/A equation by$(1+i)^n$ to obtain:

$$(1+i)^n P = A\,[(1+i)^n - 1]\,/\,i \quad \text{which yields}$$
$$F = A\,[(1+i)^n - 1]\,/\,i \quad \text{Uniform Series Compound Amount formula}$$

Solving this equation for A:

$$A = F\,[i\,/(1+i)^n - 1] \quad \text{Uniform Series Sinking Fund formula}$$

In functional notation, the uniform series factors are:

Compound Amount(F/A, i, n)

Sinking Fund(A/F, i, n)

Capital Recovery(A/P, i, n)

Present Worth(P/A, i, n)

EXAMPLE 5

If \$100 is deposited at the end of each year in a savings account that pays 6% interest per year, how much will be in the account at the end of five years?

Solution

$A = \$100$ $\quad$ $F =$ unknown
$n = 5$ years $\quad$ $i = 6\%$

$$F = A(F/A,\ i,\ n) = 100(F/A,\ 6\%,\ 5) = 100(5.637) = \$563.70$$

EXAMPLE 6
A woman wishes to make a uniform deposit every three months to her savings account so that at the end of 10 years she will have \$10,000 in the account. If the account earns 6% annual interest, compounded quarterly, how much should she deposit each three months?

Solution

$F = \$10{,}000$ $\quad$ $A =$ unknown
$n = 40$ quarterly deposits $\quad$ $i = 1\frac{1}{2}\,\%$ per quarter year

Note that i, the interest rate per interest period, is 1½ %, and there are 40 deposits.

$$A = F(A/F,\ i,\ n) = 10{,}000(A/F,\ 1\tfrac{1}{2}\,\%,\ 40) = 10{,}000(0.0184) = \$184$$

EXAMPLE 7
An individual is considering the purchase of a used automobile. The total price is \$6200 with \$1240 as a down payment and the balance paid in 48 equal monthly payments with interest at 1% per month. The payments are due at the end of each month. Compute the monthly payment.

Solution

The amount to be repaid by the 48 monthly payments is the cost of the automobile *minus* the \$1240 down payment.

$P = \$4960$ $\quad$ $A =$ unknown
$n = 48$ monthly payments $\quad$ $i = 1\%$ per month

$$A = P(A/P,\ i,\ n) = 4960(A/P,\ 1\,\%,\ 48) = 4960(0.0263) = \$130.45$$

EXAMPLE 8
A couple sold their home. In addition to cash, they took a mortgage on the house. The mortgage will be paid off by monthly payments of \$232.50 for 10 years. The couple decides to sell the mortgage to a local bank. The bank will buy the mortgage, but requires a 1% per month interest rate on their investment. How much will the bank pay for the mortgage?

Solution

$A = \$232.50$ $\quad$ $P =$ unknown
$n = 120$ months $\quad$ $i = 1\%$ per month

$$P = A(P/A,\ i,\ n) = 232.50(P/A,\ 1\%,\ 120) = 232.50(69.701) = \$16{,}205.48$$

Arithmetic Gradient

At times one will encounter a situation where the cash flow series is not a constant amount A. Instead it is an increasing series like:

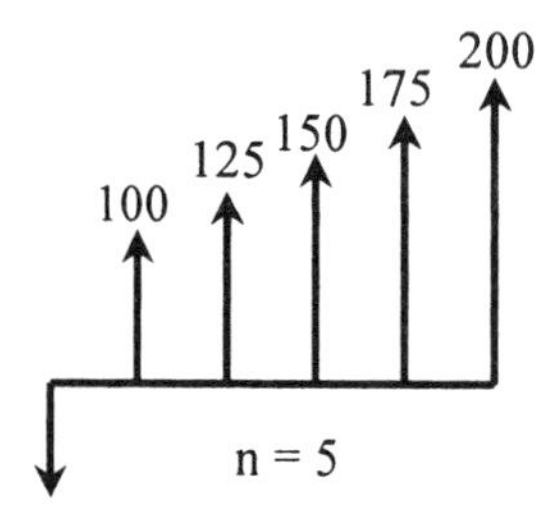

This cash flow may be resolved into two components:

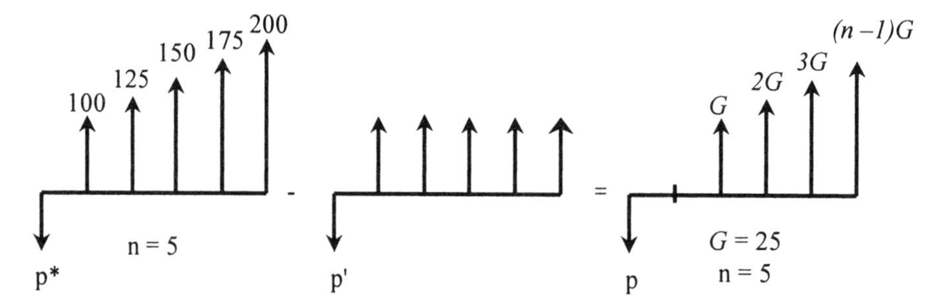

We can compute the value of P^* as equal to P' plus P. We already have an equation for P' =

$$P' = A(P/A,\ i,\ n)$$

The value for P in the right-hand diagram is:

$$P = G\left(\frac{(1+i)^n - in - 1}{i^2(1+i)^n}\right)$$

This is the Arithmetic Gradient Present Worth formula. In functional notation, the relationship is:

$$P = G(P/G,\ i,\ n)$$

EXAMPLE 9

The maintenance on a machine is expected to be \$155 at the end of the first year, and increasing \$35 each year for the following seven years. What present sum of money would need to be set aside now to pay the maintenance for the eight-year period? Assume 6 % interest.

Solution

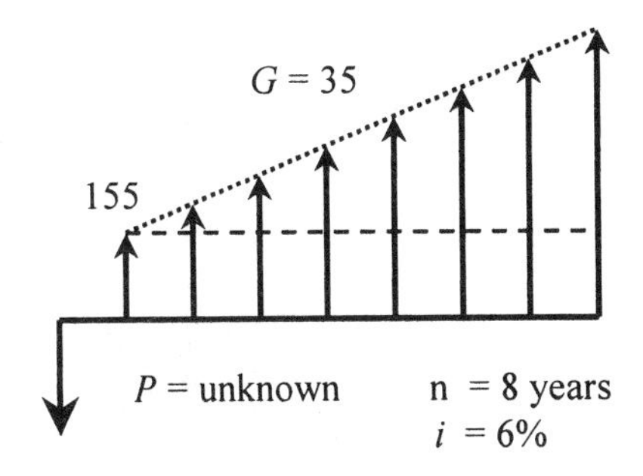

$$P = 155(P/A, 6\%, 8) + 35(P/G, 6\%, 8) = 155(6.210) + 35(19.841) = \$1656{,}99$$

In the gradient series, if instead of the present sum P, an equivalent uniform series A is desired, the problem becomes:

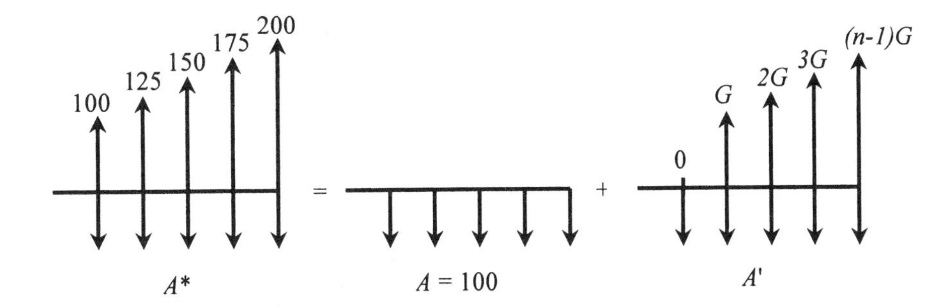

The relationship between A' and G in the right-hand diagram is:

$$A' = G\left(\frac{(1+i)^n - in - 1}{i(1+i)^n - 1}\right)$$

In functional notation, the Arithmetic Gradient (to) Uniform Series factor is:

$$A = G(A/G, i, n)$$

It is important to note carefully the diagrams for the two arithmetic gradient series factors. In both cases the first term in the arithmetic gradient series is zero and the last term is$(n - 1)G$. But we use n in the equations and functional notation. The derivations(not shown here) were done on this basis and the arithmetic gradient series Compound Interest Tables are computed this way.

EXAMPLE 10

For the situation in Example 9, we wish now to know the uniform annual maintenance cost. Compute an equivalent A for the maintenance costs.

Solution

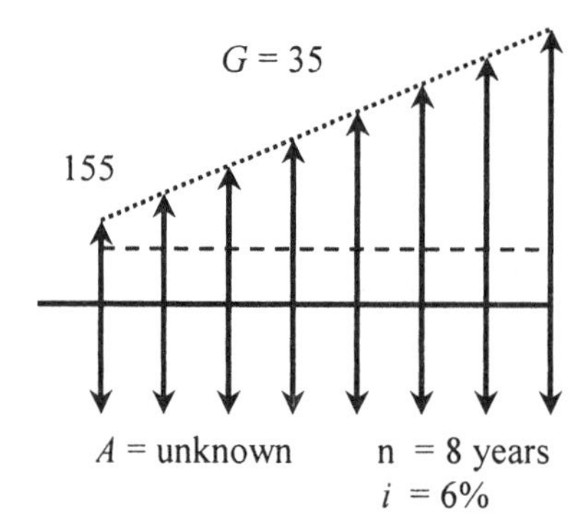

Equivalent uniform annual maintenance cost:

$$A = 155 + 35(A/G, 6\%, 8) = 155 + 35(3.195) = \$266.83$$

Geometric Gradient

The arithmetic gradient is applicable where the period-by-period change in the cash flow is a uniform amount. There are other situations where the period-by-period change is a *uniform rate, g.* A diagram of this situation is:

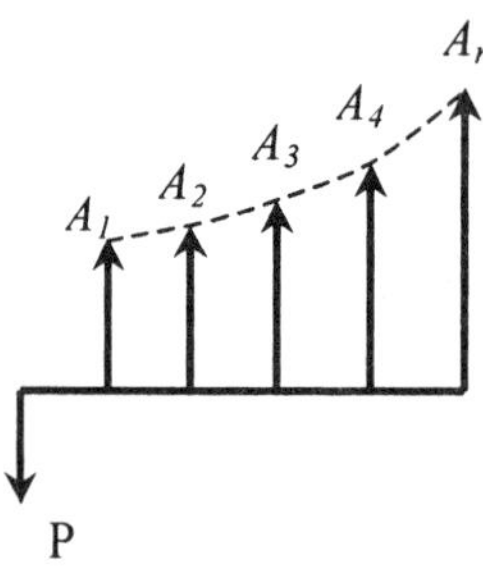

where $A_n = A_1(1 + g)^{n-1}$
g = Uniform rate of period-by-period change; the geometric gradient stated as a decimal(8% = 0.08).
A_1 = Value of A at Year 1.
A_n = Value of A at any Year n.

Geometric Series Present Worth Formulas:

When i = g, $P = A_1(n(1 + i)^{-1})$

When $i \neq$ g, $P = A_1\left(\dfrac{1-(1+g)^n(1+i)^{-n}}{i-g}\right)$

EXAMPLE 11
It is likely that airplane tickets will increase 8% in each of the next four years. The cost of a plane ticket at the end of the first year will be $180. How much money would need to be placed in a savings account now to have money to pay a student's travel home at the end of each year for the next four years? Assume the savings account pays 5% annual interest.

Solution

The problem describes a geometric gradient where $g = 8\%$ and $i = 5\ \%$.

$$P = A_1\left(\frac{1-(1+g)^n(1+i)^{-n}}{i-g}\right)$$

$$P = 180.00\left(\frac{1-(1.08)^4\ (1.05)^{-4}}{0.05-0.08}\right) \quad = \quad 180.00\left(\frac{-0.119278}{-0.03}\right) \quad = \quad \$715.67$$

Thus, $715.67 would need to be deposited now.

As a check, the problem can be solved without using the geometric gradient:

Year		*Ticket*
1	$A_1 =$	$180.00
2	$A_2 = 180.00 + 8\%(180.00) =$	194.40
3	$A_3 = 194.40 + 8\%(194.40) =$	209.95
4	$A_4 = 209.95 + 8\%(209.95) =$	226.75

$$\begin{aligned} P &= 180.00(P/F, 5\%, 1) + 194.40(P/F, 5\%, 2) + 209.95(P/F, 5\%, 3) + 226.75(P/F, 5\%, 4) \\ &= 180.00(0.9524) + 194.40(0.9070) + 209.95(0.8638) + 226.75(0.8227) \\ &= \$715.66 \end{aligned}$$

NOMINAL AND EFFECTIVE INTEREST

Nominal interest is the annual interest rate without considering the effect of any compounding.

Effective interest is the annual interest rate taking into account the effect of any compounding during the year.

Frequently an interest rate is described as an annual rate, even though the interest period may be something other than one year. A bank may pay 1½% interest on the amount in a savings account every three months. The *nominal* interest rate in this situation is 6%(4 x 1½% = 6%). But if you deposited $1000 in such an account, would you have 106%(1000) = $1060 in the account at the end of one year? The answer is no, you would have more. The amount in the account would increase as follows:

			Amount in Account
	At beginning of year	=	\$1000.00
End of 3 months:	1000.00 + 1½%(1000.00)	=	1015.00
End of 6 months:	1015.00 + 1½%(1015.00)	=	1030.23
End of 9 months:	1030.23 + 1½%(1030.23)	=	1045.68
End of one year:	1045.68 + 1½%(1045.68)	=	1061.37

The actual interest rate on the \$1000 would be the interest, \$61.37, divided by the original \$1000, or 6.137%. We call this the *effective* interest rate.

Effective interest rate $=(1 + i)^m - 1$, where

i = Interest rate per interest period;
m = Number of compoundings per year.

EXAMPLE 12

A bank charges 1½% per month on the unpaid balance for purchases made on its credit card. What nominal interest rate is it charging? What effective interest rate?

Solution

The nominal interest rate is simply the annual interest ignoring compounding, or 12(1½%) = 18%. Effective interest rate $=(1+0.015)^{12} - 1 = 0.1956 = 19.56\%$

SOLVING ECONOMIC ANALYSIS PROBLEMS

The techniques presented so far illustrate how to convert single amounts of money, and uniform or gradient series of money, into some equivalent sum at another point in time. These compound interest computations are an essential part of economic analysis problems.

The typical situation is that we have a number of alternatives and the question is, which alternative should be selected? The customary method of solution is to resolve each of the alternatives into some common form and then choose the best alternative(taking both the monetary and intangible factors into account).

Criteria

Economic analysis problems inevitably fall into one of these categories:

1. Fixed Input — The amount of money or other input resources is fixed.

Example: A project engineer has a budget of \$450,000 to overhaul a plant.

2. Fixed Output — There is a fixed task, or other output to be accomplished.

Example: A mechanical contractor has been awarded a fixed price contract to air-condition a building.

3. Neither Input nor Output Fixed — This is the general situation where neither the amount of money or other inputs, nor the amount of benefits or other outputs are fixed.

Example: A consulting engineering firm has more work available than it can handle. It is considering paying the staff for working evenings to increase the amount of design work it can perform.

There are five major methods of comparing alternatives: present worth; future worth; annual cost; rate of return; and benefit-cost ratio. These are presented in the following sections.

PRESENT WORTH

In present worth analysis, the approach is to resolve all the money consequences of an alternative into an equivalent present sum. For the three categories given above, the criteria are:

Category	*Present Worth Criterion*
Fixed Input	Maximize the Present Worth of benefits or other outputs.
Fixed Output	Minimize the Present Worth of costs or other inputs.
Neither Input nor Output Fixed	Maximize [Present Worth of benefits *minus* Present Worth of costs] or, stated another way: Maximize Net Present Worth.

Application of Present Worth

Present worth analysis is most frequently used to determine the present value of future money receipts and disbursements. We might want to know, for example, the present worth of an income producing property, like an oil well. This should provide an estimate of the price at which the property could be bought or sold.

An important restriction in the use of present worth calculations is that there must be a common analysis period when comparing alternatives. It would be incorrect, for example, to compare the present worth(PW) of cost of Pump A, expected to last 6 years, with the PW of cost of Pump B, expected to last 12 years.

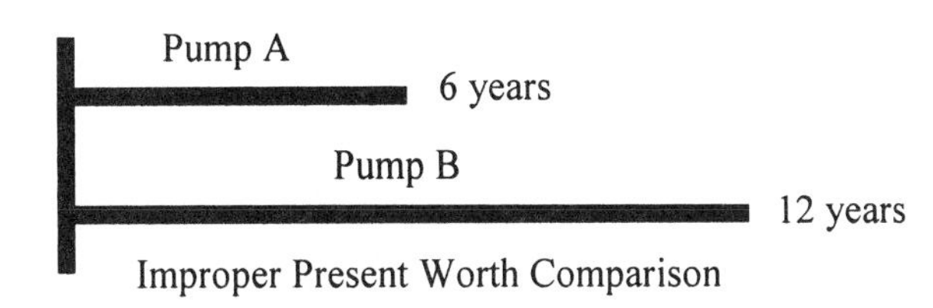

Improper Present Worth Comparison

In situations like this, the solution is either to use some other analysis technique(generally the annual cost method) or to restructure the problem so there is a common analysis period. In the example above, a customary assumption would be that a pump is needed for 12 years and that Pump A will be replaced by an identical Pump A at the end of 6 years. This gives a 12-year common analysis period.

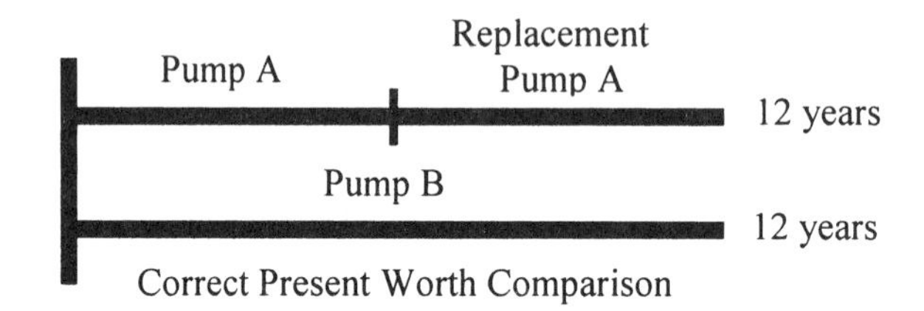

Correct Present Worth Comparison

This approach is easy to use when the different lives of the alternatives have a practical least common multiple life. When this is not true(for example, life of J equals 7 years and the life of *K* equals 11 years), some assumptions must be made to select a suitable common analysis period, or the present worth method should not be used.

EXAMPLE 13

Machine X has an initial cost of $10,000, annual maintenance of $500 per year, and no salvage value at the end of its four-year useful life. Machine Y costs $20,000. The fast year there is no maintenance cost. The second year, maintenance is $100, and increases $100 per year in subsequent years. The machine has an anticipated $5000 salvage value at the end of its 12-year useful life. If interest is 8%, which machine should be selected?

Solution

The analysis period is not stated in the, problem. Therefore we select the least common multiple of the lives, or 12 years, as the analysis period.

Present Worth of Cost of 12 years of Machine X

$$= 10{,}000 + 10{,}000(P/F, 8\%, 4) + 10{,}000(P/F, 8\%, 8) + 500(P/A, 8\%, 12)$$
$$= 10{,}000 + 10{,}000(0.7350) + 10{,}000(0.5403) + 500(7.536)$$
$$= \$26{,}521$$

Present Worth of Cost of 12 years of Machine Y

$$= 20{,}000 + 100(P/G, 8\%, 12) - 5000(P/F, 8\%, 12)$$
$$= 20{,}000 + 100(34.634) - 5000(0.3971)$$
$$= \$21{,}478$$

Choose Machine Y with its smaller PW of Cost.

EXAMPLE 14

Two alternatives have the following cash flows:

	Alternative	
Year	*A*	*B*
0	-$2000	-$2800
1	+800	+1100
2	+800	+1100
3	+800	+1100

At a 5% interest rate, which alternative should be selected?

Solution

Solving by Present Worth analysis:

Net Present Worth(NPW) = PW of benefits - PW of cost

$$\text{NPW}_A = 800(P/A, 5\%, 3) - 2000 = 800(2.723) - 2000 = +178.40$$

$$\text{NPW}_B = 1100(P/A, 5\%, 3) - 2800 = 1100(2.723) - 2800 = +195.30$$

To maximize NPW, choose Alternative *B*.

Capitalized Cost

In the special situation where the analysis period is infinite($n = \infty$), an analysis of the present worth of cost is called *capitalized cost.* There are a few public projects where the analysis period is infinity. Other examples would be permanent endowments and cemetery perpetual care.

When n equals infinity, a present sum P will accrue interest of Pi for every future interest period. For the principal sum P to continue undiminished(an essential requirement for n equal to infinity), the end-of-period sum A that can be disbursed is Pi.

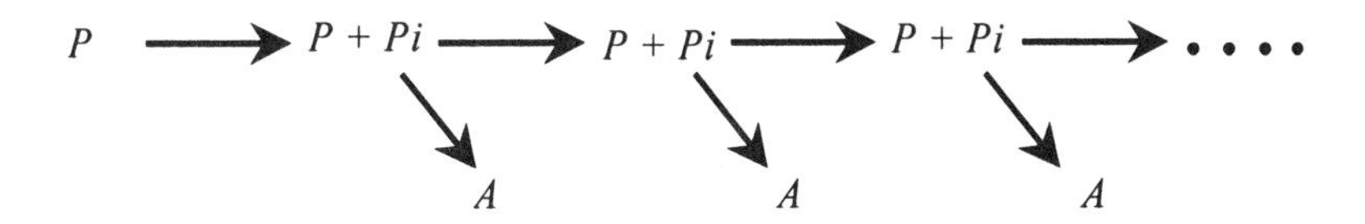

When $n = \infty$, the fundamental relationship between P, A, and i is:

$$A = Pi$$

Some form of this equation is used whenever there is a problem with an infinite analysis period.

EXAMPLE 15

In his will, a man wishes to establish a perpetual trust to provide for the maintenance of a small local park. If the annual maintenance is $7500 per year and the trust account can earn 5% interest, how much money must be set aside in the trust?

Solution

When $n = \infty$, $A = Pi$ or $P = A/i$

Capitalized cost $P = A/i = \$7500/0.05 = \$150{,}000$

FUTURE WORTH

In present worth analysis, the comparison is made in terms of the equivalent present costs and benefits. But the analysis need not be made at the present time. It could be made at any point in time: past, present, or future. Although the numerical calculations may look different, the decision is unaffected by the point in time selected. Of course, there are situations where we do want to know what the future situation will be if we take some particular course of action now. When an analysis is made based on some future point in time, it is called future worth analysis.

Category	*Future Worth Criterion*
Fixed Input	Maximize the Future Worth of benefits or other outputs.
Fixed Output	Minimize the Future Worth of costs or other inputs.
Neither Input nor Output Fixed	Maximize [Future Worth of benefits *minus* Future Worth of costs] or, stated another way: Maximize Net Future Worth.

EXAMPLE 16

Two alternatives have the following cash flows:

	Alternative	
Year	*A*	*B*
0	-$2000	-$2800
1	+800	+1100
2	+800	+1100
3	+800	+1100

At a 5% interest rate, which alternative should be selected?

Solution

In Example 14, this problem was solved by Present Worth analysis at Year 0. Here it will be solved by Future Worth analysis at the end of Year 3.

Net Future Worth(NFW) = FW of benefits - FW of cost

$$\begin{aligned} NFW_A &= 800(F/A, 5\%, 3) - 2000(F/P, 5\%, 3) \\ &= 800(3.152) - 2000(1.158) \\ &= +205.60 \end{aligned}$$

$$\begin{aligned} NPW_B &= 1100(P/A, 5\%, 3) - 2800(F/P, 5\%, 3) \\ &= 1100(3.152) - 2800(1.158) \\ &= +224.80 \end{aligned}$$

To maximize NFW, choose Alternative *B*.

ANNUAL COST

The annual cost method is more accurately described as the method of Equivalent Uniform Annual Cost(EUAC) or, where the computation is of benefits, the method of Equivalent Uniform Annual Benefits(EUAB).

For each of the three possible categories of problems, there is an annual cost criterion for economic efficiency.

Category	*Annual Cost Criterion*
Fixed Input	Maximize the Equivalent Uniform Annual Benefits. That is, maximize EUAB.
Fixed Output	Minimize the Equivalent Uniform Annual Cost. That is, minimize EUAC.
Neither Input nor Output Fixed	Maximize [EUAB - EUAC]

Application of Annual Cost Analysis

In the section on present worth, we pointed out that the present worth method requires that there be a common analysis period for all alternatives. This same restriction does not apply in all annual cost calculations, but it is important to understand the circumstances that justify comparing alternatives with different service lives.

Frequently an analysis is to provide for a more or less continuing requirement. One might need to pump water from a well, for example, as a continuing requirement. Regardless of whether the pump has a useful service life of 6 years or 12 years, we would select the one whose annual cost is a minimum. And this would still be the case if the pump useful lives were the more troublesome 7 and 11 years, respectively. Thus, if we can assume a continuing need for an item, an annual cost comparison among alternatives of differing service lives is valid.

The underlying assumption made in these situations is that when the shorter-lived alternative has reached the end of its useful life, it can be replaced with an identical item with identical costs, and so forth. This means the EUAC of the initial alternative is equal to the EUAC for the continuing series of replacements.

If, on the other hand, there is a specific requirement in some situation to pump water for 10 years, then each pump must be evaluated to see what costs will be incurred during the analysis period and what salvage value, if any, may be recovered at the end of the analysis period. The annual cost comparison needs to consider the actual circumstances of the situation.

Problems presented on examinations are often readily solved by the annual cost method. And the underlying "continuing requirement" is often present, so that an annual cost comparison of unequal-lived alternatives is an appropriate method of analysis.

EXAMPLE 17

Consider the following alternatives:

	A	*B*
First cost	\$5000	\$10,000
Annual maintenance	500	200
End-of-useful-life salvage value	600	1000
Useful life	5 years	15 years

Based on an 8% interest rate, which alternative should be selected?

Solution

Assuming both alternatives perform the same task and there is a continuing requirement, the goal is to minimize EUAC.

Alternative A;

$$\text{EUAC} = 5000(A/P, 8\%, 5) + 500 - 600(A/F, 8\%, 5)$$
$$= 5000(0.2505) + 500 - 600(0.1705) = \$1650$$

Alternative B:

$$\text{EUAC} = 10{,}000(A/P, 8\%, 15) + 200 - 1000(A/F, 8\%, 15)$$
$$= 10{,}000(0.1168) + 200 - 1000(0.0368) = \$1331$$

To minimize EUAC, select Alternative *B*.

RATE OF RETURN

A typical situation is a cash flow representing the costs and benefits. The rate of return may be defined as the interest rate where

PW of cost = PW of benefits,

EUAC = EUAB,

or PW of cost - PW of benefits = 0.

EXAMPLE 18

Compute the rate of return for the investment represented by the following cash flow:

Year	*Cash Flow*
0	-$595
1	+250
2	+200
3	+150
4	+100
5	+ 50

Solution

This declining arithmetic gradient series may be separated into two cash flows for which compound interest factors are available:

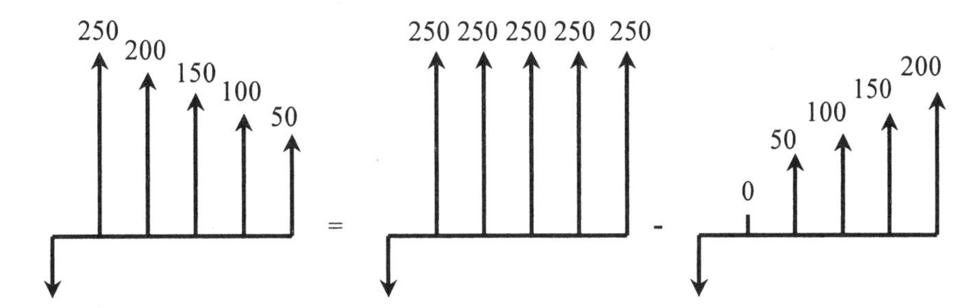

Note that the gradient series factors are based on an *increasing* gradient. Here, subtracting an increasing arithmetic gradient, as indicated by the diagram, solves the declining cash flow.

PW of cost - PW of benefits = 0

$595 - [250(P/A, i, 5) - 50(P/G, i, 5)] = 0$

Try $i = 10\%$:

$595 - [250(3.791) - 50(6.862)] = -9.65$

Try $i = 12\%$:

$595 - [250(3.605) - 50(6.397)] = +13.60$

The rate of return is between 10% and 12%. It may be computed more accurately by linear interpolation:

$$\text{Rate of return} = 10\% + (2\%)\left(\frac{9.65 - 0}{13.60 - 9.65}\right) = 10.83\%$$

Application of Rate of Return with Two Alternatives

Compute the incremental rate of return on the cash flow representing the difference between the two alternatives. Since we want to look at increments of *investment,* the cash flow for the difference between the alternatives is computed by taking the higher initial-cost alternative *minus* the lower initial-cost alternative. If the incremental rate of return is greater than or equal to the predetermined minimum attractive rate of return(MARR), choose the higher-cost alternative; otherwise, choose the lower-cost alternative.

EXAMPLE 19

Two alternatives have the following cash flows:

	Alternative	
Year	*A*	*B*
0	-$2000	-$2800
1	+800	+1100
2	+800	+1100
3	+800	+1100

If 5% is considered the minimum attractive rate of return(MARR), which alternative should be selected?

Solution

These two alternatives were previously examined in Examples 14 and 16 by present worth and future worth analysis. This time, the alternatives will be resolved using rate of return analysis.

Note that the problem statement specifies a 5% minimum attractive rate of return(MARR), while Examples 14 and 16 referred to a 5% interest rate. These are really two different ways of saying the same thing: the minimum acceptable time value of money is 5 %.

First, tabulate the cash flow that represents the increment of investment between the alternatives. Taking the higher initial-cost alternative minus the lower initial-cost alternative does this:

	Alternative		*Difference between Alternatives*
Year	*A*	*B*	*B - A*
0	-$2000	-$2800	-$800
1	+800	+1100	+300
2	+800	+1100	+300
3	+800	+1100	+300

Then compute the rate of return on the increment of investment represented by the difference between the alternatives:

$$\begin{aligned} \text{PW of cost} &= \text{PW of benefits} \\ 800 &= 300(P/A, \text{i}, 3) \\ (\text{P/A}, i, 3) &= 800/300 \\ &= 2.67 \\ i &\approx 6.1\% \end{aligned}$$

Since the incremental rate of return exceeds the 5 % MARR, the increment of investment is desirable.

Choose the higher-cost Alternative *B*.

Before leaving this example problem, one should note something that relates to the rates of return on Alternative *A* and on Alternative *B*. These rates of return, if computed, are:

	Rate of Return
Alternative *A*	9.7%
Alternative *B*	8.7%

The correct answer to this problem has been shown to be Alternative *B*, and this is true even though Alternative *A* has a higher rate of return. The higher-cost alternative may be thought of as the lower cost alternative, plus the increment of investment between them. Looked at this way, the higher-cost Alternative *B* is equal to the desirable lower-cost Alternative *A* plus the desirable differences between the alternatives.

The important conclusion is that computing the rate of return for each alternative does not provide the basis for choosing between alternatives. Instead, incremental analysis is required.

EXAMPLE 20

Consider the following:

	Alternative	
Year	*A*	*B*
0	-$200.0	-$131.0
1	+77.6	+48.1
2	+77.6	+48.1
3	+77.6	+48.1

If the minimum attractive rate of return(MARR) is 10%, which alternative should be selected?

Solution

To examine the increment of investment between the alternatives, we will examine the higher initial cost alternative minus the lower initial-cost alternative, or *A* - *B*.

	Alternative		*Increment*
Year	*A*	*B*	*A* - *B*
0	-$200.0	-$131.0	-$69.0
1	+77.6	+48.1	+29.5
2	+77.6	+48.1	+29.5
3	+77.6	+48.1	+29.5

Solve for the incremental rate of return:

$$\text{PW of cost} = \text{PW of benefits}$$
$$69.0 = 29.5(P/A, i, 3)$$
$$(P/A, i, 3) = 69.0/29.5$$
$$=2.339$$

From Compound Interest Tables, the incremental rate of return is between 12% and 15%. This is a desirable increment of investment hence we select the higher initial-cost Alternative *A*.

Application of Rate of Return with Three or More Alternatives

When there are three or more mutually exclusive alternatives, one must proceed following the same general logic presented for two alternatives. The components of incremental analysis are:

1. Compute the rate of return for each alternative. Reject any alternative where the rate of return is less than the given MARR.(This step is not essential, but helps to immediately identify unacceptable alternatives.)

2. Rank the remaining alternatives in their order of increasing initial cost.

3. Examine the increment of investment between the two lowest-cost alternatives as described for the two-alternative problem. Select the best of the two alternatives and reject the other one.

4. Take the preferred alternative from Step 3. Consider the next higher initial-cost alternative and proceed with another two-alternative comparison.

5. Continue until all alternatives have been examined and the best of the multiple alternatives has been identified.

EXAMPLE 21
Consider the following:

	Alternative	
Year	*A*	*B*
0	-$200.0	-$131.0
1	+77.6	+48.1
2	+77.6	+48.1
3	+77.6	+48.1

If the minimum attractive rate of return(MARR) is 10%, which alternative, if any, should be selected?

Solution

One should carefully note that this is a three-alternative problem where the alternatives are *A, B,* and "*Do Nothing*".

In this solution we will skip Step 1. Reorganize the problem by placing the alternatives in order of increasing initial cost:

Year	*Do Nothing*	*B*	*A*
0	0	-$131.0	-$200.0
1	0	+48.1	+77.6
2	0	+48.1	+77.6
3	0	+48.1	+77.6

Examine the "*B - Do Nothing*" increment of investment:

Year	*B*	-	*Do Nothing*		
0	-$131.0	-	0	=	-$131.0
1	+48.1	-	0	=	+48.1
2	+48.1	-	0	=	+48.1
3	+48.1	-	0	=	+48.1

Solve for the incremental rate of return:

PW of cost = PW of benefits

$131.0 = 48.1(P/A, i, 3)$

$(P/A, i, 3) = 131.0/48.1$

$= 2.723$.

From Compound Interest Tables, the incremental rate of return = 5%. Since the incremental rate of return is less than 10%, the *B - Do Nothing* increment is not desirable. Reject Alternative B.

Next, consider the increment of investment between the two remaining alternatives:

Year	*A*	-	*Do Nothing*		
0	-$200.0	-	0	=	-$131.0
1	+77.6	-	0	=	+77.6
2	+77.6	-	0	=	+77.6
3	+77.6	-	0	=	+77.6

Solve for the incremental rate of return:

PW of cost = PW of benefits

$200.0 = 77.6(P/A, i, 3)$

$(P/A, i, 3) = 200/77.6$

$= 2.577$

From Compound Interest Tables, the incremental rate of return is 8%. Since the rate of return on the *A - Do Nothing* increment of investment is less than the desired 10%, reject the increment by rejecting Alternative *A*. We select the remaining alternative: Do nothing!

If you have not already done so, you should go back to Example 20 and see how the slightly changed wording of the problem radically altered it. Example 20 required the choice between two undesirable alternatives. Example 21 adds the Do-nothing alternative that is superior to *A* or *B*.

EXAMPLE 22

Consider four mutually exclusive alternatives:

	Alternative			
	A	*B*	*C*	*D*
Initial Cost	$400.0	$100.0	$200.0	$500.0
Uniform Annual Benefit	100.9	27.7	46.2	125.2

Each alternative has a five-year useful life and no salvage value. If the minimum attractive rate of return(MARR) is 6%, which alternative should be selected?

Solution

Mutually exclusive is where selecting one alternative precludes selecting any of the other alternatives. This is the typical "textbook" situation. The solution will follow the several steps in incremental analysis.

1. The rate of return is computed for the four alternatives.

Alternative	*A*	*B*	*C*	*D*
Computed rate of return	8.3%	11.9%	5%	8%

Since Alternative C has a rate of return less than the MARR, it may be eliminated from further consideration.

2. Rank the remaining alternatives in order of increasing initial cost and examine the increment between the two lowest cost alternatives.

	Alternative		
	B	*A*	*D*
Initial Cost	$100.0	$400.0	$500.0
Uniform Annual Benefit	27.7	100.9	125.2

	A - B
Δ Initial Cost	$300.0
Δ Uniform Annual Benefit	73.2
Δ Computed Δ rate of return	7%

Since the incremental rate of return exceeds the 6% MARR, the increment of investment is desirable. Alternative *A* is the better alternative.

3. Take the preferred alternative from the previous step and consider the next higher-cost alternative. Do another two-alternative comparison.

	D - *A*
Δ Initial Cost	$100.0
Δ Uniform Annual Benefit	24.3
Δ Computed Δ rate of return	6.9%

The incremental rate of return exceeds MARR, hence the increment is desirable. Alternative *D* is preferred over Alternative *A*.

Conclusion: Select Alternative *D*. Note that once again the alternative with the highest rate of return(Alt. *B*) is not the proper choice.

BENEFIT-COST RATIO

Generally in public works and governmental economic analyses, the dominant method of analysis is called benefit-cost ratio. It is simply the ratio of benefits divided by costs, taking into account the time value of money.

$$B/C = \frac{\textbf{PW of benefits}}{\textbf{PW of cost}} = \frac{\textbf{Equivalent Uniform Annual Benefits}}{\textbf{Equivalent Uniform Annual Cost}}$$

For a given interest rate, a *B/C* ratio ≥ 1 reflects an acceptable project. The method of analysis using *B/C* ratio is parallel to that of rate of return analysis. The same kind of incremental analysis is required.

Application of B/C Ratio for Two Alternatives

Compute the incremental B / C ratio for the cash flow representing the increment of investment between the higher initial-cost alternative and the lower initial-cost alternative. If this incremental *B/C* ratio is ≥ 1, choose the higher-cost alternative; otherwise, choose the lower-cost alternative.

Application of B/C Ratio for Three or More Alternatives

Follow the procedure used for rate of return, except that the test is whether or not the incremental B/C ratio is ≥ 1.

EXAMPLE 23

Solve Example 22 using Benefit-Cost analysis. Consider four mutually exclusive alternatives:

	Alternative			
	A	*B*	*C*	*D*
Initial Cost	$400.0	$100.0	$200.0	$500.0
Uniform Annual Benefit	100.9	27.7	46.2	125.2

Each alternative has a five-year useful life and no salvage value. Based on a 6% interest rate, which alternative should be selected?

Solution

1. *B/C* ratio computed for the alternatives:

Alt. *A* $$B/C = \frac{\textbf{PW of benefits}}{\textbf{PW of cost}} = \frac{100.9(P/A, 6\%, 5)}{400} = 1.06$$

Alt. *B* $$B/C = \frac{\textbf{PW of benefits}}{\textbf{PW of cost}} = \frac{27.7(P/A, 6\%, 5)}{100} = 1.17$$

Alt. *C* $$B/C = \frac{\textbf{PW of benefits}}{\textbf{PW of cost}} = \frac{46.2(P/A, 6\%, 5)}{200} = 0.97$$

Alt. *D* $$B/C = \frac{\textbf{PW of benefits}}{\textbf{PW of cost}} = \frac{125.2(P/A, 6\%, 5)}{500} = 1.05$$

Alternative C with a *B/C* ratio less than 1 is eliminated.

2. Rank the remaining alternatives in order of increasing initial cost and examine the increment of investment between the two lowest cost alternatives.

	Alternative		
	B	*A*	*D*
Initial Cost	$100.0	$400.0	$500.0
Uniform Annual Benefit	27.7	100.9	125.2

	A - B
Δ Initial Cost	$300.0
Δ Uniform Annual Benefit	73.2

$$\text{Incremental } B/C \text{ ratio} = \frac{73.2(P/A, 6\%, 5)}{300} = 1.03$$

The incremental *B/C* ratio exceeds 1.0 hence the increment is desirable. Alternative *A* is preferred over *B*.

3. Take the preferred alternative from the previous step and consider the next higher-cost alternative. Do another two-alternative comparison.

	D - *A*
Δ Initial Cost	$100.0
Δ Uniform Annual Benefit	24.3

$$\text{Incremental } B/C \text{ ratio} = \frac{24.3(P/A, 6\%, 5)}{100} = 1.02$$

The incremental *B*/*C* ratio exceeds 1.0 hence Alternative *D* is preferred over *A*.

Conclusion: Select Alternative *D*.

BREAKEVEN ANALYSIS

In business, "breakeven" is defined as the point where income just covers the associated costs. In engineering economics, the breakeven point is more precisely defined as the point where two alternatives are equivalent.

EXAMPLE 24

A city is considering a new $50,000 snowplow. The new machine will operate at a savings of $600 per day, compared to the equipment presently being used. Assume the minimum attractive rate of return(interest rate) is 12% and the machine's life is 10 years with zero resale value at that time. How many days per year must the machine be used to make the investment economical?

Solution

This breakeven problem may be readily solved by annual cost computations. We will set the equivalent uniform annual cost of the snowplow equal to its annual benefit, and solve for the required annual utilization.

Let X = breakeven point = days of operation per year.

$$\text{EUAC} = \text{EUAB}$$

$$50{,}000(A/P, 12\%, 10) = 600\text{ X}$$

$$X = \frac{50{,}000(0.1770)}{600}$$

$$= 14.7 \text{ days/year}$$

DEPRECIATION

Depreciation of capital equipment is an important component of many after-tax economic analyses. For this reason, one must understand the fundamentals of depreciation accounting.

Depreciation is defined in its accounting sense, as the systematic allocation of the cost of a capital asset over its useful life. Book *value* is defined as the original cost of an asset, minus the accumulated depreciation of the asset.

In computing a schedule of depreciation charges, three items are considered:

1. Cost of the property, P;
2. Depreciable life in years, n;
3. Salvage value of the property at the end of its depreciable life, S.

Straight Line Depreciation

$$\text{Depreciation charge in any year} = \frac{P - S}{n}$$

Sum-Of-Years-Digits Depreciation

$$\text{Depreciation charge in any year} = \frac{\text{Remaining Depreciable Life at Beginning of Year}}{\text{Sum of Years Digits for Total Useful Life}}(P - S)$$

$$\text{where Sum Of Years Digits} = 1 + 2 + 3 + \ldots\ldots + n = \frac{n}{2}(n + 1)$$

Double Declining Balance Depreciation

$$\text{Depreciation charge in any year} = \frac{2}{n}(P \text{ - Depreciation charges to date})$$

Modified Accelerated Cost Recovery System Depreciation

Modified accelerated cost recovery system(MACRS) depreciation is based on a property class life that is generally less than the actual useful life of the property and on zero salvage value. The varying depreciation percentage to use must be read from a table(based on declining balance with conversion to straight line). Unless one knows the proper MACRS property class, the depreciation charge in any year cannot be computed.

The following tables are used with MACRS depreciation. Table 1 categorizes assets by recovery period. Table 2 provides the percentage depreciation allowed each year.

Depreciation charge in any year $n = P$(percentage depreciation allowance)

Recovery Periods for MACRS

Recovery Period	Description of Assets Included
3-Year	Tractors for over-the-road tractor/trailer use and special tools such as dies and jigs; ADR< 4 years
5-Year	Cars, buses, trucks, computers, office machinery, construction equipment, and R&D equipment; 4 years # ADR < 10 years
7-Year	Office furniture, most manufacturing equipment, mining equipment, And items not otherwise classified, 10 years # ADR < 16 years
10-Year	Marine vessels, petroleum refining equipment, single-purpose agricultural structures, trees and vines that bear nuts or fruits; 16 years # ADR < 20 years
15-Year	Roads, shrubbery, wharves, steam and electric generation and Distribution systems, and municipal wastewater treatment facilities; 20 years # ADR < 25 years
20-Year	Farm buildings and municipal sewers; ADR ∃ 25 years
27.5-Year	Residential rental property
31.5-Year	Non-residential real property purchased on or before 5/12/93
39-Year	Non-residential real property purchased on or after 5/13/93

Note: This table is used to find the recovery period for *all* items under MACRS.

MACRS Percentages

	Recovery Period					
Year	3-year	5-year	7-year	10-year	15-year	20-year
1	33.33	20.00	14.29	10.00	5.00	3.750
2	44.45	32.00	24.49	18.00	9.50	7.219
3	14.81	19.20	17.49	14.40	8.55	6.677
4	8.41	11.52	12.49	11.52	7.70	6.177
5		11.52	8.93	9.22	6.93	5.713
6		5.76	8.92	7.37	6.23	5.285
7			8.93	6.55	5.90	4.888
8			4.46	6.55	5.90	4.522
9				6.56	5.91	4.462
10				6.55	5.90	4.461
11				3.28	5.91	4.462
12					5.90	4.461
13					5.91	4.462
14					5.90	4.461
15					5.91	4.462
16					2.95	4.461
17						4.462
18						4.461
19						4.462
20						4.461
21						2.231

EXAMPLE 25

A piece of construction machinery costs \$5000 and has an anticipated \$1000 salvage value at the end of its five year depreciable life. Compute the depreciation schedule for the machinery by:

(a) Straight line depreciation;
(b) Sum-of-years-digits depreciation;
(c) Double declining balance depreciation.
(d) MACRS

Solution

$$\text{Straight line depreciation} = \frac{P-S}{n} = \frac{5000-1000}{5} = \$800$$

Sum - of - years - digits depreciation:

$$\text{Sum-of-years-digits} = \frac{n}{2}(n+1) = \frac{5}{2}(6) = 15$$

$$\text{1st year depreciation} = \frac{5}{15}(5000 - 1000) = \$1333$$

$$\text{2nd year depreciation} = \frac{4}{15}(5000 - 1000) = 1067$$

$$\text{3rd year depreciation} = \frac{3}{15}(5000 - 1000) = 800$$

$$\text{4th year depreciation} = \frac{2}{15}(5000 - 1000) = 533$$

$$\text{5th year depreciation} = \frac{1}{15}(5000 - 1000) = \underline{267}$$

$$\$4000$$

Double declining balance depreciation:

$$\text{1st year depreciation} = \frac{2}{5}(5000 - 0) = \$2000$$

$$\text{2nd year depreciation} = \frac{2}{5}(5000 - 2000) = 1200$$

$$\text{3rd year depreciation} = \frac{2}{5}(5000 - 3200) = 720$$

$$\text{4th year depreciation} = \frac{2}{5}(5000 - 3920) = 432$$

$$\text{5th year depreciation} = \frac{2}{5}(5000 - 4352) = \underline{259}$$

$$\$4611$$

Since the problem specifies a \$1000 salvage value, the total depreciation may not exceed \$4000. The double declining balance depreciation must be stopped in the 4th year when it totals \$4000.

MACRS

Since the asset is specified as construction equipment the 5-year recovery period is determined. Note that the depreciation will be over a six year period. This is because only ½ year depreciation is allowed in the first year. The remaining ½ year is recovered in the last year. If an asset is disposed of mid-life, than only ½ of the disposal year's depreciation is taken. Also note that the salvage value is not used in the depreciation calculations.

1st year depreciation	= .2000(5000) =	$1000
2nd year depreciation	= .3200(5000) =	1600
3rd year depreciation	= .1920(5000) =	960
4th year depreciation	= .1152(5000) =	576
5th year depreciation	= .1152(5000) =	576
6th year depreciation	= .0576(5000) =	288
		$5000

The depreciation schedules computed by the four methods are as follows:

Year	*StL*	*SOYD*	*DDB*	*MACRS*
1	$800	$1333	$2000	$1000
2	800	1067	1200	1600
3	800	800	720	960
4	800	533	80	576
5	800	267	0	576
6				288

INCOME TAXES

Income taxes represent another of the various kinds of disbursements encountered in an economic analysis. The starting point in an after-tax computation is the before-tax cash flow. Generally, the before-tax cash flow contains three types of entries:

1. Disbursements of money to purchase capital assets. These expenditures create no direct tax consequence for they are the exchange of one asset(cash) for another(capital equipment).

2. Periodic receipts and/or disbursements representing operating income and/or expenses. These increase or decrease the year-by-year tax liability of the firm.

3. Receipts of money from the sale of capital assets, usually in the form of a salvage value when the equipment is removed. The tax consequence depends on the relationship between the residual value of the asset and its book value(cost - depreciation taken).

Residual Value is	Tax Consequence
less than current book value	loss on sale
equals current book value	no loss & no recapture
exceeds current book value	recaptured depreciation
exceeds initial book value	recaptured depreciation & capital gain

After the before-tax cash flow, the next step is to compute the depreciation schedule for any capital assets. Next, taxable income is the taxable component of the before-tax cash flow minus the depreciation. Then, the income tax is the taxable income times the appropriate tax rate. Finally, the after-tax cash flow is the before-tax cash flow adjusted for income taxes.

To organize these data, it is customary to arrange them in the form of a cash flow table, as follows:

Year	*Before-tax cash flow*	*Depreciation*	*Taxable income*	*Income taxes*	*After-tax cash flow*
0	-	-	-	-	-
1	-	-	-	-	-

EXAMPLE 26

A corporation expects to receive $32,000 each year for 15 years from the sale of a product. There will be an initial investment of $150,000. Manufacturing and sales expenses will be $8067 per year. Assume straight-line depreciation, a 15-year useful life and no salvage value. Use a 46% income tax rate. Determine the projected after-tax rate of return.

Solution

$$\text{Straight line depreciation} = \frac{P-S}{n} = \frac{150{,}000-0}{15} = \$10{,}000 \text{ per year}$$

Year	*Before-tax cash flow*	*Depreciation*	*Taxable income*	*Income taxes*	*After-tax cash flow*
0	-150,000				-150,000
1	+23,933	10,000	13,933	-6409	+17,524
2	+23,933	10,000	13,933	-6409	+17,524
.	.	.	.	.	.
.	.	.	.	.	.
.	.	.	.	.	.
15	+23,933	10,000	13,933	-6409	+17,524

Take the after-tax cash flow and compute the rate of return at which PW of cost equals PW of benefits.

$$150{,}000 = 17{,}524(P/A, i, 15)$$
$$(P/A, i, 15) = 150{,}000/17{,}524 = 8.559$$

From Compound Interest Tables, $i = 8\%$.

INFLATION

Inflation is a decrease in the buying power of the dollar (or peso, yen, etc.). More generally it is thought of as the increase in the level of prices. Inflation rates are usually estimated as having a constant rate of change. This type of cash flow is a geometric gradient.

Future cash flows are generally estimated in constant-value terms. The assumption is made that the inflation rates for various cash flows will match the economy's inflation rate. This allows for the use of a real interest rate.

Another approach is the market interest rate. The market rate includes both the time value of money and an estimate of current and predicted inflation.

For exact calculation of real and market interest rates the following formula is used.

$$(1 + \text{Market rate}) = (1 + i)(1 + f)$$

where i = the real interest rate

f = the inflation rate

EXAMPLE 27

Determine the market interest rate if the inflation rate is estimated to be 2.5% and the time value of money (the real interest rate) is 6%.

Solution

$$(1 + \text{Market rate}) = (1 + .06)(1 + .025)$$
$$(1 + \text{Market rate}) = 1.0865$$
$$\text{Market rate} = 0.0865 = 8.65\%$$

Uncertainty and Probability

It is unrealistic to consider future cash flows to be known exactly. Cash flows, useful lifes, salvage values, etc. can all be expressed using probability distributions. Then measures of expected or average return and of risk can be used in the decision making process.

Probabilities can be derived from the following:

1. Historical trends or data.
2. Mathematical models (Probability distributions).
3. Subjective estimates

Probabilities must satisfy the following:

1. $0 \le P \le 1$.
2. The sum of all probabilities must = 1.

Expected values for cash flows, useful lifes, salvage values, etc. can be found by using a weighted average calculation.

EXAMPLE 28

The following table summarizes estimated annual benefits, annual costs, and end-of-life value for an asset under consideration. Determine the expected value for each of the three.

	State and Associated Probability			
	p = .20	p = .40	p = .25	p = .15
Annual Benefits	3,000	6,500	8,000	11,500
Annual Costs	1,500	2,800	5,000	5,750
End-of-Life-Value	5,000	7,500	9,000	10,000

Solution

E(Annual Benefits) = .20(3,000) + .40(6,500) + .25(8,000) + .15(11,500)
= $6,925.00

E(Annual Costs) = .20(1,500) + .40(2,800) + .25(5,000) + .15(5,750)
= $3,532.50

E(End-of-Life-Value) = .20(5,000) + .40(7,500) + .25(9,000) + .15(10,000)
= $7,750.00

Chapter 1

Making Economic Decisions

1-1

Many engineers earn high salaries for creating profits for their employers, and then find themselves at retirement time insufficiently prepared financially. This may be because in college courses emphasis seldom is placed on using engineering economics for the direct personal benefit of the engineer. Among the goals of every engineer should be assuring that adequate funds will be available for anticipated personal needs at retirement.

A realistic goal of retiring at age 60 with a personal net worth in excess of one million dollars can be accomplished by several methods. A recent independent study ranked the probability of success of the following methods of personal wealth accumulation. Discuss and decide the ranking order of the following five methods.

(a) Purchase as many lottery tickets as possible with money saved from salary.
(b) Place money saved from salary in a bank savings account.
(c) Place all money saved from a salary in a money market account.
(d) Invest saved money into rental properties and spend evenings, weekends and vacations repairing and managing.
(e) Invest all money saved into stock market securities, and study investments 10 to 15 hours per week.

Solution

Independent studies can be misleading. If Julia McNeese of Roseburg, Oregon were asked to rank wealth accumulation methods, (a) would head her list. Julia recently won one million dollars in a Canadian lottery. A workaholic with handyman talent might select (d) as his Number 1 choice. Lots of people have become millionaires by investing in real estate. The important thing is to learn about the many investment vehicles available and then choose the one or the several most suitable for you.

1-2

A food processor is considering the development of a new line of product. Depending on the quality of raw material, he can expect different yields process-wise, and the quality of the final products will also change considerably. The product development department has identified three alternatives, and produced them in a pilot scale. The marketing department has used those samples for surveys to estimate potential sales and pricing strategies. The three alternatives would use existing equipment, but different process conditions and specifications, and they are summarized as follows. Indicate which alternative seems to be the best according to the estimated data, if the objective is to maximize total profit per year.

	Alternative		
	1	2	3
Lbs of raw material A per unit of product	0.05	0.07	0.075
Lbs of raw material B per unit of product	0.19	0.18	0.26
Lbs of raw material C per unit of product	0.14	0.12	0.17
Other processing costs ($/unit product)	$0.16	$0.24	$0.23
Expected wholesale price ($/unit product)	0.95	1.05	1.25
Projected volume of sales (units of product)	1,000,000	1,250,000	800,000

Cost of raw material A $3.45/lb

Cost of raw material B $1.07/lb

Cost of raw material C $1.88/lb

Solution

		Alternative		
		1	2	3
Cost of Raw Material. A ($/unit product)	.05 x 3.45 =	0.1725	0.2415	0.2587
Cost of Raw Material B ($/unit product)	.19 x 1.07 =	0.2033	0.1926	0.2782
Cost of Raw Material C ($/unit product)	.14 x 1. 88 =	0.2632	0.2256	0.3196
Other processing costs ($/unit product)		$0.16	$0.24	$0.23
Total Cost ($/unit product)		0.799	0.8997	1.0865
Wholesale price ($/unit product)		0.95	1.05	1.25
Profit per unit		0.151	0.1503	0.1635
Projected sales (units of product)		1,000,000	1,250,000	800,000
Projected profits		151,000	187,875	130,800

Therefore, choose alternative 2.

1-3

Consider the previous problem. When asked about the precision of the given figures, the marketing department indicated the actual sales results could change plus/minus 10% from the forecast. Similarly, product development indicated the actual production costs may vary 3% from the pilot-based calculations. Is your choice of the best alternatives the same as you found in the previous problem?

Solution

Alternative 1:

Cost	Profit
max.=1.03 x .799 =.82297	min. =.95-.82297=.12703
min. =0.97 x .799 =.77503	max.=.95-.77503=.17497
Sales	**Sales x Profit = TOTAL PROFIT**
max.=1.1 x 1,000,000 =1,100,000	max.=1,100,000 x .17497=$192,467
min. =0.9 x 1,000,000 = 900,000	min. = 900,000 x .12703=$114,327

Alternative 2:
max. total profit = (1,250,000 x 1.1)(1.05 -.8997 x 0.97) = $243,775
min total profit = (1,250,000 x 0.9)(1.05 -.8997 x 1.03) = $138,723

Alternative 3:
max.total profit = (800,000 x 1.1)(1.25 - 1.0865 x 0.97) = $172.564
min. total profit = (800,000 x 0.9)(1.25 - 1.0865 x 1.03) = $94,252

Although alternative 2 still gives the largest profits, the data's precision is not good enough to tell for sure that it will actually be the best choice since the max. profits for 1 & 3 are larger than the min. profits of 2.

1-4

Car A initially costs $500 more than Car B, but it consumes 0.04 gallons/mile versus 0.05 gallons/mile for B. Both last 8 years and B's salvage value is $100 smaller than A's. Fuel costs $1.70 per gallon. Other things being equal, beyond how many miles of use per year (X) does A become preferable to B?

Solution

$$-500 + 100 + (.05 - .04)(1.70)(8)X = 0$$
$$-400 + 0.136X = 0$$
$$X = 400/0.136$$
$$= 2{,}941 \text{ miles/year}$$

1-5

The following letter was a reply from Benjamin Franklin to Joseph Priestley, a friend of Franklin's. Priestley had been invited to become the librarian for the Earl of Shelburne and had asked for Franklin's advice. What engineering economy principle does Franklin suggest Priestley use to aid in making his decision?

London, September 19, 1772
Dear Sir:

In the affair of so much importance to you wherein you ask my advice, I cannot, for want of sufficient premises, advise you what to determine, but if you please I will tell you how. When these difficult cases occur, they are difficult chiefly because while we have them under consideration, all the reasons Pro and Con are not present to the mind at the same time; but sometimes one set present themselves, and at other times another, the first being out of sight. Hence the various purposes or inclination that alternately prevail, and the uncertainty that perplexes us.

To get over this, my way is to divide a half a sheet of paper by a line into two columns; writing over the one PRO and over the other CON. Then during three or four days' consideration I put down under the different heads short hints of the different motives that at different times occur to me, for or against the measure. When I have thus got them all together in one view, I endeavour to estimate their respective weights; and where I find two (one on each side) that seem equal, I strike them both out. If I find a reason Pro equal to some two reasons Con, I strike out the three. If I judge some two reasons Con equal to three reasons Pro, I strike out the five; and thus proceeding I find at length where the balance lies; and if after a day or two of further consideration, nothing new that is of importance occurs on either side, I come to a determination accordingly. And though the weight of the reasons cannot be taken with the precision of algebraic quantities, yet when each is thus considered separately and comparatively and the whole lies before me, I think I can judge better, and am less likely to make a rash step; and in fact I have found great advantage from this kind of equation in what may be called moral or prudential algebra.

Wishing sincerely that you may determine for the best, I am ever, my dear friend, your most affectionately...

s/Ben Franklin

Solution

Decisions should be based on the differences between the alternatives. Here the alternatives are taking the job (Pro) and not taking the job (Con).

1-6

Assume that you are employed as an engineer for Wreckall Engineering, Inc., a firm specializing in demolition of high-rise buildings. The firm has won a bid to tear down a 30-story building in a heavily developed downtown area. The crane owned by the company only reaches to 29 stories. Your boss asks you to perform an economic analysis of buying a new crane to complete the job. How would you handle the analysis?

Solution

The important point of this problem is to realize that your boss may not have recognized what the true problem is in this case. To buy a new crane is only one alternative, and quite likely not the best alternative.

Other alternatives: extension on current crane
ramp for current crane
rent a crane to remove top story
explosive demolition
.... etc.

If this is a fixed output project (e.g., fixed fee for demolishing building) we want to minimize costs. Weigh alternatives using economic criteria to choose the best alternative.

1-7

The total cost of a building (TC) is given by

$$TC = \left(200 + 80X = 2X^2\right)A \quad \text{where } X = \text{Number of floors}$$

$$A = \text{Floor area in } ft^2 / \text{floor}$$

If the total number of square feet required is 10^6, what is the optimal (minimum cost) number of floors?

Solution

$$TC = \left(200 + 80X + 2X^2\right)\left(10^6 / x\right)$$

$$\frac{dTC}{dx} = \left(10^6\right)\left(\frac{-200}{x^2} + 2\right) = 0$$

$$X^* = \sqrt{\frac{200}{2}} = \sqrt{100} = 10 \text{ floors}$$

1-8

By saving and investing, wisely or luckily or both, Helen finds she has accumulated $400,000 in savings while her salaried position is providing her with $40,000 per year, including benefits, and after income taxes and other deductions.

Helen's salaried position is demanding and allows her little free time, but the desire to pursue other interests has become very strong. What would be your advice to her if you were asked?

Solution

First, Helen should decide what annual income she needs to provide herself with the things she wants. Depending on her age, she might be able to live on the interest income (maybe 10% x $400,000 = $40,000), or a combination of interest and principal. The important thing that Helen should realize is that it may be possible for her to lead a more fulfilling lifestyle if she is fully aware of the time value of money. There are many people with large sums of money in bank checking accounts (drawing no interest) because they can write "free" checks.

1-9

Charles belongs to a square dance club that meets twice each month and has quarterly dues of $9.00 per person. The club moved its meeting place to a location with increased cost. To offset the cost each member agrees to pay 50 cents each time they attend the meeting. Later the treasurer suggests that the quarterly dues be increased to $12.00 per person as an alternative to the meeting charge. Discuss the consequences of the proposal. Do you think the club members would agree to the proposal?

Solution

The members who attend regularly would pay the same amount with the new dues as with the older method of $9.00 plus 50 cents per meeting. Many would like the added advantage of covering their quarterly expenses in one check. The members who attend infrequently would pay more by the new method and might oppose the action.

Since the people who attend infrequently are in the minority in this club, the members voted to approve the proposal.

1-10

Sam decides to buy a cattle ranch and leave the big city rat race. He locates an attractive 500-acre spread in Montana for $1,000 per acre that includes a house, a barn, and other improvements. Sam's studies indicate that he can run 200 cow-calf pairs and be able to market 180 500-pound calves per year. Sam, being rather thorough in his investigation, determines that he will need to purchase an additional $95,000 worth of machinery. He expects that supplemental feeds, medications and veterinary bills will be about $50 per cow per year. Property taxes are $4000 per year, and machinery upkeep and repairs are expected to run $3,000 per year.

If interest is 10% and Sam would like a net salary of $10,000 per year, how much will he have to get for each 500-pound calf?

Solution

Land Cost :	$500 Acre x $1,000/Acre	= $500,000
Machinery:	Lump sum	= 95,000
Total Fixed Cost		$595,000

Assume lands and machinery to have a very long life
At 10% Annual Cost = (.10)($595,000) = $59,500
Other Annual Costs:

Feeds, medications, vet bills $50 x 200 =	$ 10,000
Property taxes	4,000
Upkeep & Repairs	3,000
Salary	10,000
Total Annual Cost	86,500

Net sale price of each calf would have to be: $86,500/180 = $480.56

Note: If Sam were to invest his $595,000 in a suitable investment vehicle yielding 10% interest his salary would be almost six times greater and he could go fishing instead of punching cows.

Chapter 2

Engineering Costs and Cost Estimating

2-1

A small community outside of Atlanta Georgia is planning to construct a new fire station. As currently planned it will have 7,000 square-feet under roof. The area cost factor is 86% of the 144-city average. The estimated cost per square foot for a typical 3,500 square facility is \$98. Based on economies of scale a size adjustment factor of 95% can be used. Estimate the cost of the construction. Assume a cost growth factor of 1.364.

Solution

Estimated cost = 98(7,000)(.95)(.86)(1.364)
= \$764,470

2-2

In 2000 a new 21-kW power substation was built in Gibson county Tennessee for 1.4 million dollars. Weakley county, a neighboring county, is planning on building a similar though smaller substation in 2003. An 18-kW substation is planned for Weakley county. The inflation rate between 2001 and 2003 has averaged 1.5% per year. If the power sizing exponent is .85 for this type of facility what is the estimated cost of construction?

Solution

Cost of the 21-kW substation in 2003 dollars = $1{,}400{,}000(1.015)^3$ = \$1,463,950

$$C_x = C_k(S_x/S_k)^n$$

$$C_{21} = C_{18}(18/21)^{.85} = 1{,}463{,}950(.8772)$$
$$= \$1{,}284{,}177$$

2-3

The time required to produce the first gizmo is 1500 blips. Determine the time required to produce the 450th gizmo if the learning-curve coefficient is .85.

Solution

$T_i = T_1\Theta^{(\ln i/\ln 2)}$

$T_{450} = 1500(.85)^{(\ln 450/\ln 2)} = 358.1$ blips

2-4

Four operations are required to produce a certain product produced by ABC Manufacturing. Using information presented in the table below, determine the labor cost of producing the 1000 piece.

	Time required for 1st piece	Learning curve coefficient	Labor cost per hour
Operation 1	1 hour 15 minutes	.90	$ 8.50
Operation 2	2 hours	.82	12.00
Operation 3	2 hours 45 minutes	.98	7.75
Operation 4	4 hours 10 minutes	.74	10.50

Solution

$T_i = T_1\Theta^{(\ln i/\ln 2)}$

Operation 1

$T_{1000} = 75(.90)^{(\ln 1000/\ln 2)} = 26.25$ minutes
Cost = 26.25/60 x 8.50 = $3.72

Operation 2

$T_{1000} = 120(.82)^{(\ln 1000/\ln 2)} = 16.61$ minutes
Cost = 16.61/60 x 12.00 = $3.32

Operation 3

$T_{1000} = 165(.98)^{(\ln 1000/\ln 2)} = 134.91$ minutes
Cost = 134.91/60 x 7.75 = $17.43

Operation 4

$T_{1000} = 250(.74)^{(\ln 1000/\ln 2)} = 12.44$ minutes
Cost = 12.44/60 x 10.50 = $2.18

Total cost = 3.72 + 3.32 + 17.43 + 2.18 = $26.65

2-5

American Petroleum (AP) recently completed construction on a large refinery in Texas. The final construction cost was \$17,500,000. The refinery covers a total of 340 acres. The Expansion and Acquisition Department at AP is currently working on plans for a new refinery for the panhandle of Oklahoma. The anticipated size is approximately 260 acres. If the power sizing exponent is .67 for this type of facility what is the estimated cost of construction?

Solution

$$C_x = C_k(S_x/S_k)^n$$

$$C_{260} = C_{340}(260/340)^{.67} = 17{,}500{,}00(.83549)$$
$$= \$14{,}621{,}075$$

2-6

A new training program at Arid Industries is intended to lower the learning curve coefficient of a certain molding operation that currently costs \$95.50/hour. The current coefficient is .87 and the program hopes to lower the coefficient by 10%. Assuming the time to mold the first product is 8 hours. What cost savings can be realized when the 2000th piece is produced if the program is successful?

Solution

$$T_i = T_1\Theta^{(\ln i/\ln 2)}$$

Without the training program:

$$T_{2000} = 8(.87)^{(\ln 2000/\ln 2)} = 1.74 \text{ hours}$$

With the training program:

$$T_{2000} = 8(.783)^{(\ln 2000/\ln 2)} = .547 \text{ hours}$$

Cost savings = (1.74 - .547)(95.50) = \$113.93

2-7

The following data is has been provided by XYZ Manufacturing concerning one of their most popular products. Estimate the selling price per unit.

Labor	= 12.8 hours at \$18.75/hour
Factory overhead	= 92% of labor
Material costs	= \$65.10
Packing cost	= 10% of materials
Sales commission	= 10% of selling price
Profit	= 22% of selling price

Solution

Labor Cost	=	12.8 x 18.75	= $240.00
Factory overhead	=	92% of labor	= 220.80
Material cost	=		= 65.10
Packing Cost	=	10% of material costs	= 6.51
			$532.41

Let X be the selling price

$$0.10X + 0.22X + 532.41 = X$$
$$0.68X = 532.41$$
$$X = 532.41/0.68 = \$782.96$$

Chapter 3

Interest and Equivalence

3-1
Solve for the unknown value. Be sure to show your work.

$P = 1{,}000 \quad i = 12\% \quad n = 5 \quad F = ?$

Solution

$F = 1{,}000(F/P, 12\%, 5) = \$1{,}762.00$

3-2
If you had $1,000 now and invested it at 6%, how much would it be worth 12 years from now?

$F = 1{,}000(F/P, 6\%, 12) = \$2{,}012.00$

3-3
Mr. Beach deposited $200,000 in the Lawrence National Bank. If the bank pays 8% interest, how much will he have in the account at the end of 10 years?

Solution

$F = 200{,}000(F/P, 8\%, 10) = \$413{,}800$

3-4
If you can make 6% interest on your money, how much is $1000 paid to you 12 years in the future worth to you now?

Solution

$P = 1{,}000(P/F, 6\%, 12) = = \$497.00$

3-5

Downtown is experiencing an explosive population growth of 18% per year. At the end of 2002 the population was 16,000. If the growth rate continues unabated, at the end of how many years will the population have passed 75,000?

Solution

Use i = 18% to represent the growth rate.

$$75{,}000 = 16{,}000(F/P, 18\%, n)$$
$$(F/P, 18\%, n) = 75{,}000/16{,}000 = 4.6875$$

From tables n is 10

Note that population would not have passed 75,000 after 9 years.

3-6

If the interest rate is 6% compounded quarterly, how long (number of quarters) does it take to earn $100 interest on an initial deposit of $300?

Solution

$$i = 6\%/4 = 1\tfrac{1}{2}\%$$

$$400 = 300(F/P, 1\tfrac{1}{2}\%, n)$$
$$(F/P, 1\tfrac{1}{2}\%, n) = 400/300 = 1.333$$

From tables n = 20 quarters

3-7

A man decides to put $100 per month into an account paying 12% compounded monthly. Without using formulas or factors (that is, use only basic concepts) determine how much (to the penny) will be in the account immediately after the fourth deposit.

Solution

Month	Beginning Balance	Interest @ 1%	Deposit	Ending Balance
1	$ 0.00	0.00	$100	$100.00
2	100.00	1.00	100	201.00
3	201.00	2.01	100	303.01
4	303.01	3.03	100	Ans→ 406.04

3-8

Write the functional notation of the future \$500 to determine P, and evaluate.

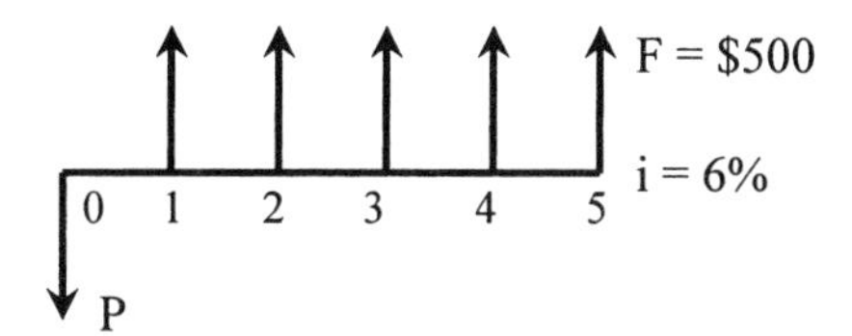

Solution

$$P = F(P/F, 6\%, 5) = \$500(0.7473) = \$373.65$$

3-9

Draw a diagram that represents (P/F, i%, 3).

Solution

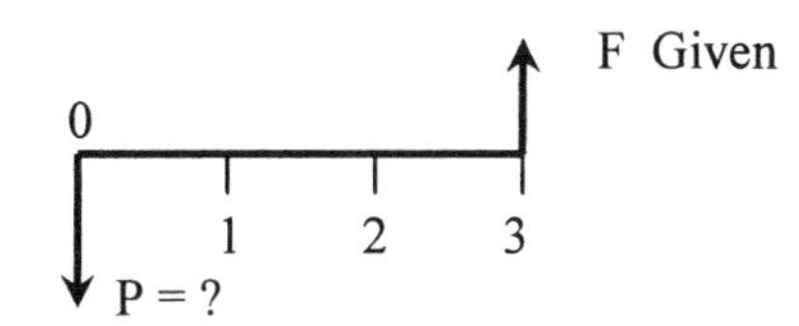

3-10

On July 1 and September 1, Abby placed \$2,000 into an account paying 3% compounded monthly. How much was in the account on October 1?

Solution

$$F = 2{,}000(1 + .0025)^3 + 2{,}000(1 + .0025)^1 = \$4{,}020.04$$

OR

Month	Beginning Balance	Interest @ ½%	Ending Balance	Deposit
July	\$2,000.00	\$10.00	\$2,010.00	0
August	2,010.00	10.05	2,020.05	\$2,000.00
September	4,020.05	20.10	4,040.15	0
October	Answer→ 4,040.15			

Chapter 4

More Interest Formulas

4-1
The effective interest rate is 19.56%. If there are 12 compounding periods per year, what is the nominal interest rate?

Solution

$i_{eff} = (1 + (r/m))^m - 1 \Rightarrow r/m = (1 + i_{eff})^{1/m} - 1 = (1.1956)^{1/12} - 1 = 1.5\%$

$r = 12 \times 1.5 = 18\%$

4-2
A continuously compounded loan has what nominal interest rate if the effective interest rate is 25%? Select one of the five choices below.

(a) $e^{1.25}$
(b) $e^{0.25}$
(c) $\log_e (1.25)$
(d) $\log_e (0.25)$
(e) Other (specify) ____________________

Solution

$$e^r - 1 = .25$$
$$e^r = 1.25$$
$$\log_e (e^r) = \log_e (1.25)$$
$$r = \log_e (1.25)$$

Choose c.

4-3

A continuously compounded loan has what effective interest rate if the nominal interest rate is 25%? Select one of the five choices below.

(a) $e^{1.25}$
(b) $e^{0.25}$
(c) $\log_e(1.25)$
(d) $\log_e(0.25)$
(e) Other (specify) ______________________

Solution

$$i_{eff} = e^{.25} - 1$$

Choose e.

4-4

Given: A situation where the annual interest rate is 5%. When continuous compounding is used, rather than monthly compounding, the nominal interest rate. (Select One)

(a) Increases
(b) Remains the same
(c) Decreases

Solution

The answer is b: Remains the same

4-5

A journalist for a small town newspaper has a weekly column in which he answers questions from the local populace on various financial matters. Below is one such question. Assume you are the journalist and respond to this inquiry. Be brief and specific.

Q: "I put $10,000 in a 12% six month savings certificate. When it matured, I expected to receive interest of $1200 - 12% of $10,000. Instead, I received only $600. Why? Current six month certificates now pay 9% interest. If I put $10,000 in a new six month certificate, I assume (based on my previous experience) I'll get only $450 interest - one half of 9%. Wouldn't my money earn more interest if I deposit it in a savings account paying 5 1/2 percent?"

Solution

A: The questioner is confused about nominal vs. effective interest rates. The 9% and 12% rates are nominal annual interest rates compounded semi-annually. The effective semi-annual interest rates are 0.09/2 = 0.045 and 0.12/2 = 0.06 hence the interest earned in 6 months would be 0.045 (10,000) = \$450 and 0.06 (10,000) = \$600. The corresponding effective annual interest rates are

$$(1.045)^2 - 1 = 0.092 \text{ and } (1.06)^2 - 1 = 0.1236$$

The interest rate advertised on the savings account typically is also a nominal annual rate. Most accounts pay interest quarterly or continuously, thus the effective annual interest rate would be either

$$(1.01375)^4 - 1 = 0.05614 \quad \text{or} \quad e^{0.055} - 1 = 0.05654$$

neither of which is as good as the six month certificates.

4-6

A drug dealer will sell goods to his regular customers for \$20 immediately or \$22 if the payment is deferred one week. What nominal annual interest rate is the dealer receiving?

Solution

$i = 2/20 = 10\%$

$r = 52 \times 10\% = 520\%$

4-7

A local bank is advertising that they pay savers 6% compounded monthly, yielding an effective annual rate of 6.168%. If \$2,000 is placed in savings now and no withdrawals are made, how much interest (to the penny) will be earned in one year?

Solution

Interest = Effective annual rate x principal = 0.06168 x 2,000 = \$123.36

Monthly compounding is irrelevant when the effective rate is known.

4-8

A young engineer wishes to buy a house but only can afford monthly payments of \$500. Thirty year loans are available at 12% interest compounded monthly. If she can make a \$5,000 downpayment, what is the price of the most expensive house that she can afford to purchase?

Solution

i 12%/12 = 1 % n = (30)(12) = 360

P* = 500(P/A, 1%, 360) = 48,609
P = 48,609 + 5000
P = $53,609

4-9

A small company borrowed $10,000 to expand the business. The entire principal of $10,000 will be repaid in 2 years but quarterly interest of $330 must be paid every three months. What nominal annual interest rate is the company paying?

Solution

Since $330 is interest only for one interest period, then i = 330/10,000 = 3.3% per quarter

r = 3.3 x 4 = 13.2% nominal annual

4-10

A store policy is to charge 1¼% interest each month on the unpaid balance.

(a) What is the nominal interest?

(b) What is the effective interest?

Solution

(a) r = mi = 12(1.25) = 15%

(b) $i_{eff} = (1 + i)^n - 1 = (1.0125)^{12} - 1 = 16.075\%$

4-11

Under what circumstances are the nominal and effective annual interest rates exactly equal; or is this never true?

Solution

The nominal interest rate equals the effective interest rate when there is yearly (annual) compounding (i.e., m = 1).

4-12

A small company borrowed \$10,000 to expand the business. The entire principal of \$10,000 will be repaid at the end of two years but quarterly interest of \$335 must be paid every three months. What nominal annual interest rate is the company paying?

Solution

$i = 335/10{,}000 = 3.35\%$

$r = i\,m = 4 \times 3.35 = 13.40\%$

4-13

E. Z. Marc received a loan of \$50 from the S.H. Ark Loan Company that he had to repay one month later with a single payment of \$60. What was the nominal annual interest rate for this loan?

Solution

Interest = \$10 in one month

$i = 10/50 = 20\%$

$r = im = 20 \times 12 = 240\%$

4-14

A local college parking enforcement bureau issues parking tickets that must be paid within one week. The person receiving the ticket may pay either \$5 immediately, or \$7 if payment is deferred one week. What nominal interest rate is implied in the arrangement?

Solution

$i = (7 - 5)/5 = 40\%$ per week

$r = i\,m = 52(40) = 2080\%$

4-15

A deposit of \$300 was made one year ago into an account paying monthly interest. If the account now has \$320.52, what was the effective annual interest rate? Give answer to 1/100 of a percent.

Solution

$i_{eff} = 20.52/300 = 6.84\%$

4-16

Which is the better investment, a fund that pays 15% compounded annually, or one that pays 14% compounded continuously?

Solution

$i = 15\%$; $n = 1$ $F = P(1 + .15)^{1} = 1.1500\,P$

$r = 14\%$ cont; $n = 1$ $F = P\,e^{.14} = 1.1503\,P$

14% compounded continuously is slightly better

4-17

For a nominal interest of 16 percent, what would the effective interest be, if interest is

(a) compounded quarterly?
(b) compounded monthly?
(c) compounded continuously?

Solution

(a) $i_{eff} = [(1.04)^{4} - 1\,](\,100) = 16.986\%$

(b) $i_{eff} = [(1.01333)^{12} - 1\,](100) = 17.222\%$

(c) $i_{eff} = e^{rn} - 1 = e^{.16(1)} - 1 = 17.35\%$

4-18

If compounding is weekly and the (one year = 12 months = 48 weeks for this problem) quarterly effective interest rate is 5%,

(a) What is the nominal annual interest rate?
(b) What is the weekly interest rate?
(c) What is the semi-annual effective interest rate?
(d) What is the effective interest rate for a two year period?

Solution

(a) $i_P = (1 + r/m)^{P} - 1 \rightarrow r = m\,\{\,(1 + i_P)^{1/P} - 1\,\}$

$i_{qr} = i_{12} = .05 = (1 + r/48)^{12} - 1$

$r = 48\,\{1.05^{1/12} - 1\,\} = 19.555\%$ per year compounded weekly.

(b) $i_{WK} = r/m = 19.555 / 48 = .4074\%$ per week

(c) $i_{SA} = i_{24} = i_P = (1 + i)^P - 1 = (1.004074)^{24} - 1 = 10.25\%$ per 1/2 year.

(d) $i_{2\ YR.} = i_{96} = i_P = (1 + i)^P - 1 = (1.004074)^{96} - 1 = 47.75\%$ per 2 years.

4-19

If the interest rate is 10% compounded continuously, what is the semi-annual effective interest rate?

Solution

$$i_{t} = e^{rt} - 1$$

$$i_{1/2} = e^{(.1)(1/2)} - 1 = e^{.05} - 1 = 1.0512711 - 1 = 5.127\% \text{ per } 1/2 \text{ year}$$

4-20

A grandfather gave his grandchild \$100 for his 10th birthday. The child's parents talked him into putting this gift into a bank account so that when he had grandchildren of his own he could give them similar gifts. The child lets this account grow for 50 years, and it has \$100000. What was the interest rate of the account?

(a) 15.0%
(b) 15.8%
(c) 14.8%
(d) 14.0%

Solution

$$\$100{,}000 = \$100(1 + i)^{50}$$
$$i = 14.8\%$$

The answer is c.

4-21

Given : Nominal interest = 9%, compounded monthly

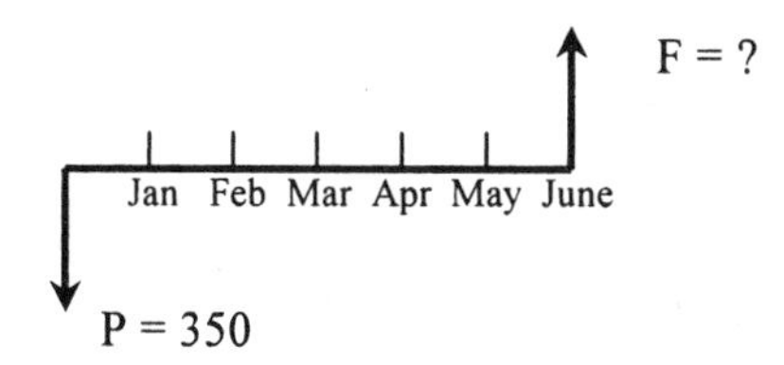

Find: (a) F
(b) i_{eff}

Solution

(a) $F = 350(F/P, .75\%, 6) = 366.10$
(b) $i_{eff} = (1 + i\,)^m - 1 = (1.0075)^{12} - 1 = 9.38\%$

4-22
Given:

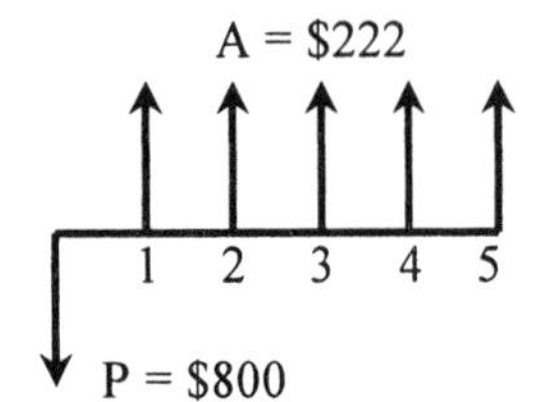

Find: i %

Solution

$$P = A(P/A, i\,\%, 5)$$
$$800 = 222(P/A, i\%, 5)$$
$$(P/A, i\,\%, 5) = 800/222$$
$$= 3.6$$

From tables $i = 12\%$

4-23
How much should Abigail invest in a fund that will pay 9%, compounded continuously, if she wishes to have $600,000 in the fund at the end of 10 years?

Solution

$r = 0.09$
$n = 10$

$$P = F(P/F, 9\%, \infty) = Fe^{-rn} = 600{,}000(0.40657) = \$243{,}941.80$$

4-24
Solve for the unknown interest rate.

$P = \$1{,}000$ $n = 10$ years $A = \$238.50$ $i = ?$

Solution

$$A = P(A/P, i\%, 10)$$
$$238.50 = 1{,}000(A/P, i\%, n)$$
$$(A/P, i\%, n) = 238.50/1{,}000 = 0.2385$$

From tables, $i = 20\%$

4-25

How much will accumulate in an Individual Retirement Account (IRA) in 15 years if $500 is deposited in the account at the end of each quarter during that time? The account earns 8% interest, compounded quarterly. What is the effective interest rate?

Solution

$i = 8/4 = 2\%$ $\quad n = (4)(15) = 60$

$F = 500\ (F/A, 3\%, 60) = \$57{,}025.50$

Effective interest rate $= (1 + .02)^4 - 1 = 8.24\%$

4-26

Solve for the unknown value. Be sure to show your work.

$P = 1{,}000 \quad i = 12\% \quad n = 5 \quad A = ?$

Solution

$A = 1{,}000(A/P, 12\%, 5) = \$277.40$

4-27

To offset the cost of buying a $75,000 house, a couple borrowed $12,500 from their parents at 6% nominal interest, compounded monthly. The loan from their parents is to be paid off in five years in equal monthly payments. The couple has saved $11,250. Their total downpayment is therefore $12,500 + 11,250 = $23,750. The balance will be mortgaged at 9% nominal interest, compounded monthly for 30 years.

Find the combined monthly payment that the couple will be making for the first five years.

Solution

Payment to parents:

$$12{,}500(A/P, \tfrac{1}{2}\%, 60) = \$241.25$$

Borrowed from bank: 75,000 - 23,750 = $51,250

Payment to bank

$$51{,}250(A/P, \tfrac{3}{4}\%, 360) = \$412.56$$

Therefore, monthly payments are $241.25 + 412.56 = $653.81

4-28

Find C if i = 12%

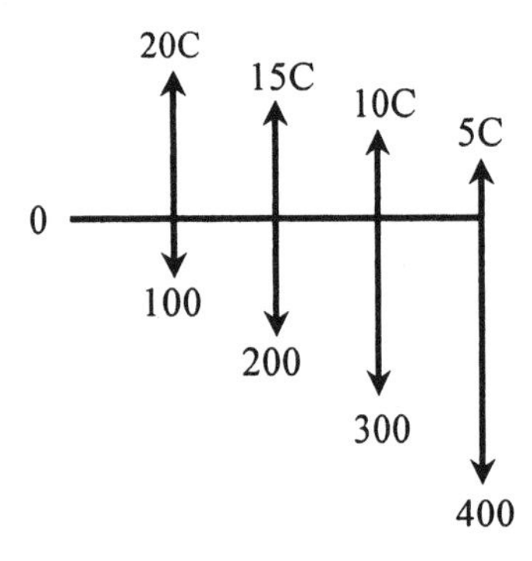

Solution

$$\begin{aligned} 20C(P/A, 12\%, 4) - 5C(P/G, 12\%, 4) &= 100(P/A, 12\%, 4) + 100(P/G, 12\%, 4) \\ 40.105C &= 716.4 \\ C &= 17.86 \end{aligned}$$

4-29

Decide whether each of the three statements below is TRUE or FALSE without referring to your book, notes, or compound interest tables.

(a) If interest is compounded quarterly, the interest period is four months.

(b) (F/A, 12%, 30) = (F/A, 1%, 360)

(c) (F/P, i %, 10) is greater than (P/F, i %, 10) for all values of i% > 0%.

Solution

(a) FALSE. If interest is compounded quarterly, each interest period is 3 months long.

(b) FALSE. If we assume, for example, we are talking about 30 years or 360 months, the (F/A, 12%, 30) does not provide monthly compounding of interest. The (F/A, 1%, 360) does. (This is a common error among beginning students.)

(c) TRUE. Since i % > 0 %, (1 + i) > 0. Thus (F/P, i %, 10) = $(1+i)^{10} > 1$ and (P/F, i %, 10) = $(1 + i)^{-10} < 1$ for all values of i % > 0%.

4-30

A company borrowed $10,000 at 12% interest. The loan was repaid according to the following schedule. Find X, the amount that will pay off the loan at the end of year 5.

Year	Amount
1	$2,000
2	2,000
3	2,000
4	2,000
5	X

Solution

10,000 = 2,000 (P/A, 12%, 4) + X(P/F, 12%, 5)
3,926 = X(.5674)
X = 3,926/.5674
= $6,919.28

4-31

How much will Thomas accumulate in a bank account if he deposits $3,000 at the end of each year for 7 years? Use interest = 5% per annum.

Solution

F = 3,000(F/A, 5%, 7) = $25,962.

4-32

You need to borrow $10,000 and the following two alternatives are available at different banks:

(a) Pay $2,571 at the end of each year for 5 years, starting at the end of the first year. (5 payments in total.)
(b) Pay $207.58 at the end of each month, for 5 years, starting at the end of the first month. (60 payments in total.)

On the basis of the interest rate being charged in each case, which alternative should you choose?

Solution

Alternative a:

$$10{,}000 = 2{,}571(P/A, i, 5)$$
$$(P/A, i, 5) = 10{,}000/2{,}571 = 3.890$$

From tables; i $\approx$ 9% (nominal = effective rate since compounded annually)

Alternative b:

$$n = 5 \times 12 = 60$$

$$10{,}000 = 207.58(P/A, i, 60)$$
$$(P/A, i, 60) = 10{,}000/\ 207.58 = 48.174$$

From tables i = .75%

The nominal annual interest rate is: $12 \times .75 = 9\%$
but the effective interest rate is: $(1 + (0.09/12))^{12} - 1 = 9.38\%$

Therefore, choose the first alternative.

4-33

Find the Uniform Equivalent for the following cash flow diagram if i = 18%. Use the appropriate gradient and uniform series factors.

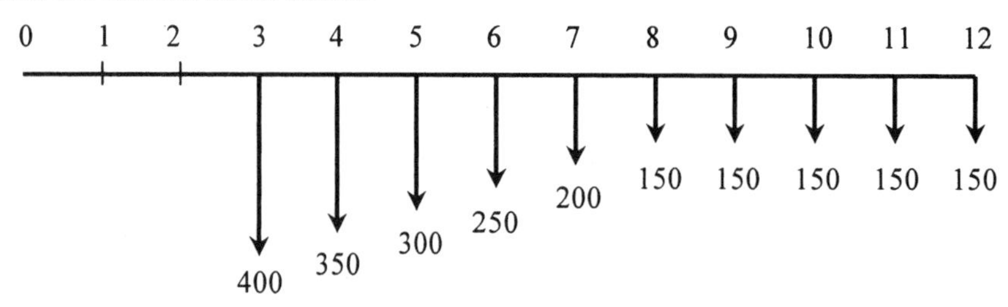

Solution

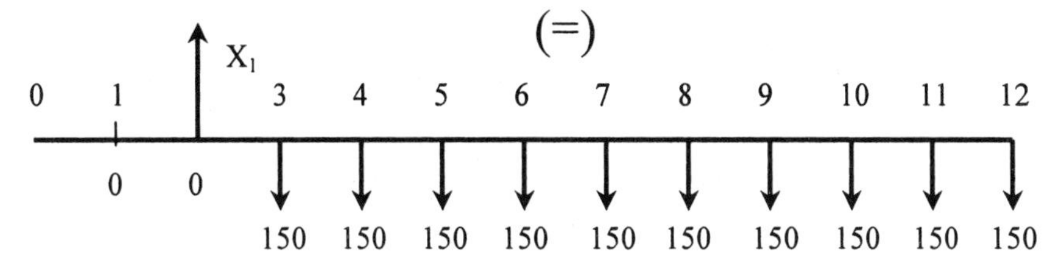

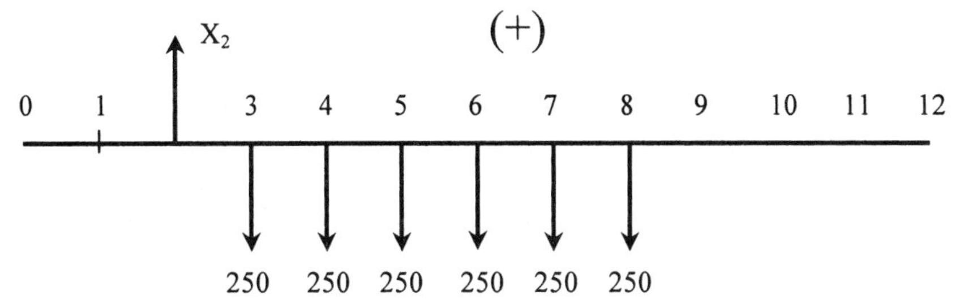

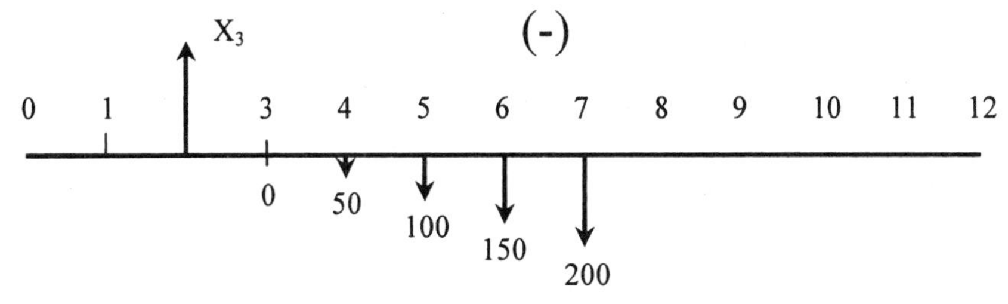

X_1 = 150 (P/A, 18%, 10) = 674.10
X_2 = 250 (P/A, 18%, 5) = 781.75
X_3 = 50 (P/G, 18%, 5) = 261.55

X = $X_1 + X_2 - X_3$ = \$1,194.30

P = 1,194.30 (P/F, 18%, 2) = 857.75
A = 857.75 (A/P, 18%, 12) = \$178.93

4-34

Below is an equation to compute an equivalent annual cash flow (EACF). Determine the values of the net cash flow series that is implied by the equation.

EACF = [-8000 - 8000(P/F, 10%,1)](A/P, 10%, 8)
+ 2000 +500(P/G, 10%, 4)(P/F, 10%, 1)(A/P, 10%, 8)
+ 750[(P/F, 10%.6) - (P/F, 10%, 8)](A/P, 10%, 8)

Time	Net Cash Flow Series
0	
1	
2	
3	
4	
5	
6	
7	
8	

Solution

Time	Net Cash Flow Series				
0	- 8,000 =	- 8,000			
1	-6,000 =	- 8,000	+ 2,000		
2	2,000 =		+ 2,000		
3	2,500 =		+ 2,000	+ 500	
4	3,000 =		+ 2,000	+ 1,000	
5	3,500 =		+ 2,000	+ 1,500	
6	2,750 =		+ 2,000		+750
7	2,000 =		+ 2,000		
8	1,250 =		+ 2,000		- 750

4-35

Using a credit card, Ben Spendthrift has just purchased a new stereo system for $975 and will be making payments of $45 per month. If the interest rate is 18% compounded monthly, how long will it take to completely pay off the stereo?

Solution

$$i = 18\%/12 = 1\tfrac{1}{2}\%$$

$$975 = 45(P/A, 1\tfrac{1}{2}\%, n)$$
$$(P/A, 1\tfrac{1}{2}\%, n) = 975/45 = 21.667$$

From tables n is between 26 and 27 months. The loan will not be completely paid off after 26 months. Therefore the payment in the 27th month will be smaller.

4-36

An engineer on the verge of retirement has accumulated savings of $100,000 that are in an account paying 6% compounded quarterly. The engineer wishes to withdraw $6000 each quarter. For how long can she withdraw the full amount?

Solution

$$i = 6\%/4 = 1\tfrac{1}{2}\%$$

$$6{,}000 = 100{,}000(A/P, 1\tfrac{1}{2}\%, n)$$
$$(A/P, 1\tfrac{1}{2}\%, n) = 0.0600$$

From tables n = 19 quarters or $4\tfrac{3}{4}$years

Note: This leaves some money in the account but not enough for a full $6,000 withdrawal

4-37

Charles puts $25 per month into an account at 9% interest for two years to be used to purchase an automobile. The car he selects then costs more than the amount in the fund. He agrees to pay $50 per month for two more years, at 12% interest, and also makes cash downpayment of $283.15. What is the cost of the automobile?

Solution

$$P = 283.15 + 25(F/A, \tfrac{3}{4}\%, 24) + 50(P/A, 1\%, 24)$$
$$= 283.15 + 654.70 + 1{,}062.15$$
$$= \$2{,}000 \leftarrow \text{ cost of auto}$$

4-38

Explain in one or two sentences why (A/P, i%, infinity) = i.

Solution

In order to have an infinitely long A series, the principal must never be reduced. For this to happen only the interest earned each period may be removed. Removing more than the interest would deplete the principal so that even less interest is available the next period.

4-39

A bank is offering a loan of $20,000 with a nominal interest rate of 12%, payable in 48 months.

(a) Calculate first the monthly payments.
(b) This bank also charges a loan fee of 4% of the amount of the loan, payable at the time of the closing of the loan (that is, at the time they give the money to the borrower). What is the effective interest rate they are charging?

Solution

(a) The monthly payments:

n = 48 ; i = 12%/12 = 1% per period (month)

20,000(A/P, 1%, 48) = $526.00

(b) Actual money received = P = 20,000 – 0.04(20,000) = $19,200

But A = 526.00; n = 48

Recalling that A = P(A/P, i, n)
526 = 19,200(A/P, i, 48)
(A/P, i, 48) = 526/19,200
= 0.02739

for i = 1¼% the A/P, factor @ n = 48 = 0.0278

for i = 1% the A/P factor @ n = 48 = 0.0263

by interpolation $i \approx 1 + ¼((.0263 - .02739)/(.0263 - .0278))$

$i \approx 1.1817\%$

the effective interest rate = $(1 + 0.011817)^{12} - 1 = 0.1514 = 15.14\%$

4-40

Find A if A = $3,000(A/P, 13.5%, ∞).

Solution

$A = P \times i$ when $n = \infty$

$A = 3{,}000 \times .135 = \$405$

4-41

Henry Fuller purchases a used automobile for $4,500. He wishes to limit his monthly payment to $100 for a period of two years. What downpayment must he make to complete the purchase if the interest rate is 9% on the loan?

Solution

$$\begin{aligned} P &= P' + A(P/A, \tfrac{3}{4}\%, 24) \\ 4{,}500 &= P' + 100(21.889) \\ P' &= 4{,}500 - 2{,}188.90 \\ &= \$2{,}437.60 \leftarrow \text{downpayment} \end{aligned}$$

4-42

Find the present equivalent of the following cash flow diagrams if i = 18%.

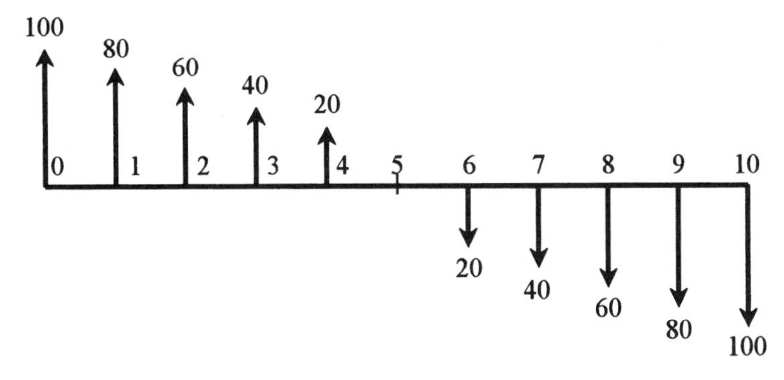

Solution

$P = 100 + 80(P/A, 18\%, 10) - 20(P/G, 18\%, 10) = \$172.48$

4-43

To start business, ECON ENGINEERING has just borrowed \$500,000 at 6%, compounded quarterly, which will be repaid by quarterly payments of \$50,000 each, with the first payment due in one year. How many quarters after the money is borrowed is the loan fully paid off?

Solution

$$i = 6/4 = 1\tfrac{1}{2}\%$$

$$500{,}000 = 50{,}000(P/A, 1\tfrac{1}{2}\%, n)(P/F, 1\tfrac{1}{2}\%3)$$
$$(P/A, 1\tfrac{1}{2}\%, n) = 500{,}000/[50{,}000(.9563)]$$
$$= 10.46$$

From tables n = 12 payments plus 3 quarters without payments equal 15 quarters before loan is fully paid off.

4-44

Given:

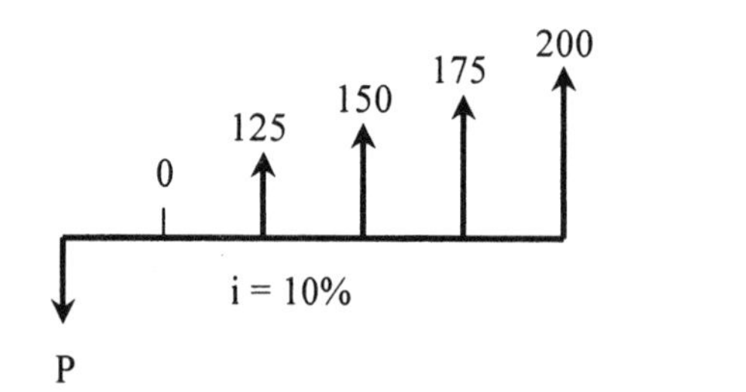

Find: P

Solution

$$P = [125(P/A, 10\%, 4) + 25(P/G, 10\%, 4)](P/F, 10\%, 1) = 459.73$$

4-45

A tractor is bought for \$125,000. What is the required payment per year to completely pay off the tractor in 20 years, assuming an interest rate of 6%?

(a) \$ 1,150
(b) \$ 5,550
(c) \$10,900
(d) \$12,750

Solution

$$A = 125{,}000(A/P, 6\%, 20)$$
$$= \$10{,}900$$

Answer is c.

Chapter 5

Present Worth Analysis

5-1

Sarah and her husband decide they will buy $1,000 worth of utility stocks beginning one year from now. Since they expect their salaries to increase, they will increase their purchases by $200 per year for the next nine years. What would the present worth of all the stocks be if they yield a uniform dividend rate of 10% throughout the investment period and the price/share remains constant?

Solution

PW of the base amount ($1,000) is: 1,000(P/A, 10%, 10) = $6,144.57

PW of the gradient is: 200(P/G, 10%, 10) = $4,578.27

Total PW = 6,144.57 + 4,578.27 = $10,722.84

5-2

Using an interest rate of 8%, what is the capitalized cost of a tunnel to transport water through the Lubbock mountain range if the first cost is $1,000,000 and the maintenance costs are expected to occur in a 6-year cycle as shown below?

End of Year:	1	2	3	4	5	6
Maintenance:	$35,000	$35,000	$35,000	$45,000	$45,000	$60,000

Solution

Capitalized Cost = PW of Cost for an infinite time period. As the initial step, compute the Equivalent Annual Maintenance Cost.

EAC = 35,000 + [10,000(F/A, 8%, 3) + 15,000](A/F, 8%, 6) = $41,468.80

For n = ∞, P = A/I

Capitalized Cost = 1,000,000 + (41,468.80/0.08) = $1,518,360.

5-3

The investment in a crane is expected to produce profit from its rental as shown below, over the next six years. Assume the salvage value is zero. What is the present worth of the investment, assuming 12% interest?

Year	Profit
1	$15,000
2	12,500
3	10,000
4	7,500
5	5,000
6	2,500

Solution

P = 15,000(P/A, 12%, 6) - 2,500(P/G, 12%, 6) = $39,340

5-4

A tax refund expected one year from now has a present worth of $3000 if i = 6 %. What is its present worth if i = 10 %?

Solution

Let x = refund value when received at the end of year 1 = 3,000(F/P, 6%, 1);
PW_x = x(P/F, 10%, 1)
Therefore the PW if i = 10% = 3,000(F/P, 6%, 1)(P/F, 10%, 1) = $2,890.94

5-5

It takes $10,000 to put on a Festival of Laughingly Absurd Works each year. Immediately before this year's FLAW, the sponsoring committee finds that it has $60,000 left in an account paying 8% interest. After this year, how many more FLAWs can be sponsored without raising more money? Think Carefully!

Solution

60,000 - 10,000 = 10,000(P/A, 8%, n)
(P/A, 8%, n) = 50,000/10,000
= 5

Therefore n = 6 which is the number of FLAWs after this year's. There will be some money left over but not enough to pay for a 7th year.

5-6
An engineer is considering buying a life insurance policy for his family. He currently owes about \$77,500 in different loans, and would like his family to have an annual available income of \$35,000 indefinitely (that is, the annual interest should amount to \$35,000 so that the original capital does not decrease).

(a) He feels he can safely assume that the family will be able to get a 4% interest rate on that capital. How much life insurance should he buy?

(b) If he now assumes the family can get a 7% interest rate, calculate again how much life insurance should he buy.

Solution

(a) If they get 4% interest rate:

$n = \infty$
$A = Pi$ or $P = A/i = 35{,}000/0.04 = 875{,}000$

Total life insurance = 77,500 + 875,000 = \$952,500

(b) If they can get 7% interest rate:

again $n = \infty$
$P = A/i = 35{,}000/0.07 = 500{,}000$

Total life insurance = 77,500 + 500,000 = \$577,500

5-7
The winner of a sweepstakes prize is given the choice of one million dollars or the guaranteed amount of \$80,000 a year for 20 years. If the value of money is taken at a 5% interest rate, which choice is better for the winner?

Solution

Alternative 1: P = \$1,000,000

Alternative 2: P = 80,000K(P/A, 5%, 20) = 81K(7.469) = \$996,960

Choose alternative 1: take \$1,000,000 now

5-8
The annual income from an apartment house is \$20,000. The annual expense is estimated to be \$2000. If the apartment house could be sold for \$100,000 at the end of 10 years, how much could you afford to pay for it now, with 10% considered a suitable interest rate?

Solution

$P = (A_{INCOME} - A_{EXPENSES})(P/A, i\,\%, n) + F_{RE\text{-}SAIE}(P/F, i\,\%, n)$
$= (20{,}000 - 2{,}000)(P/A, 10\%, 10) + 100{,}000(P/F, 10\%, 10)$
$= \$149{,}160$

5-9

A scholarship is to be established that will pay $200 per quarter at the beginning of Fall, Winter, and Spring quarters. It is estimated that a fund for this purpose will earn 10% interest, compounded quarterly. What lump sum at the beginning of Summer quarter, when deposited, will assure that the scholarship may be continued into perpetuity?

Solution

$P = 200(P/A, 2\ 1/2\ \%, 3) = 571.20$

$A' = 571.20(A/P, 2\ 1/2\ \%, 4) = 151.82$

For $n = \infty$, $P' = A' / i = 151.82 / .025 = \$6{,}073$ deposit

5-10

Your company has been presented with an opportunity to invest in a project. The facts on the project are presented below:

Investment required	$60,000,000
Salvage value after 10 years	0
Gross income expected from the project	20,000,000/yr
Operating costs:	
Labor	2,500,000/yr
Materials, licenses, insurance, etc	1,000,000/yr
Fuel and other costs	1,500,000/yr
Maintenance costs	500,000/yr

The project is expected to operate as shown for ten years. If your management expects to make 25% on its investments before taxes, would you recommend this project?

Solution

$PW = -60{,}000{,}000 + 14{,}500{,}000(P/A, 25\%, 10) = -\$8{,}220{,}500$

Reject due to negative NPW

5-11

Find the Present Equivalent of the following cash flow diagram if i = 18 %.

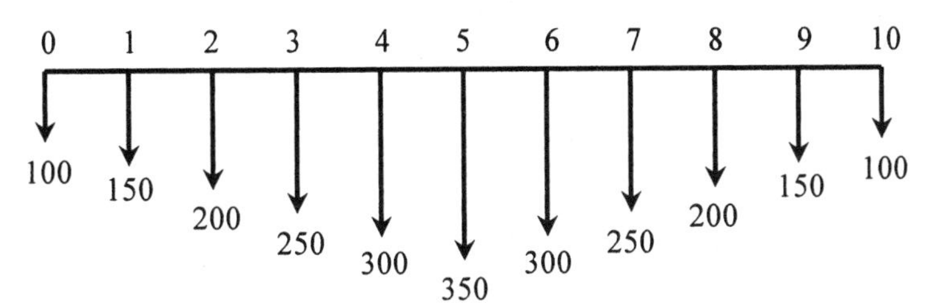

Solution

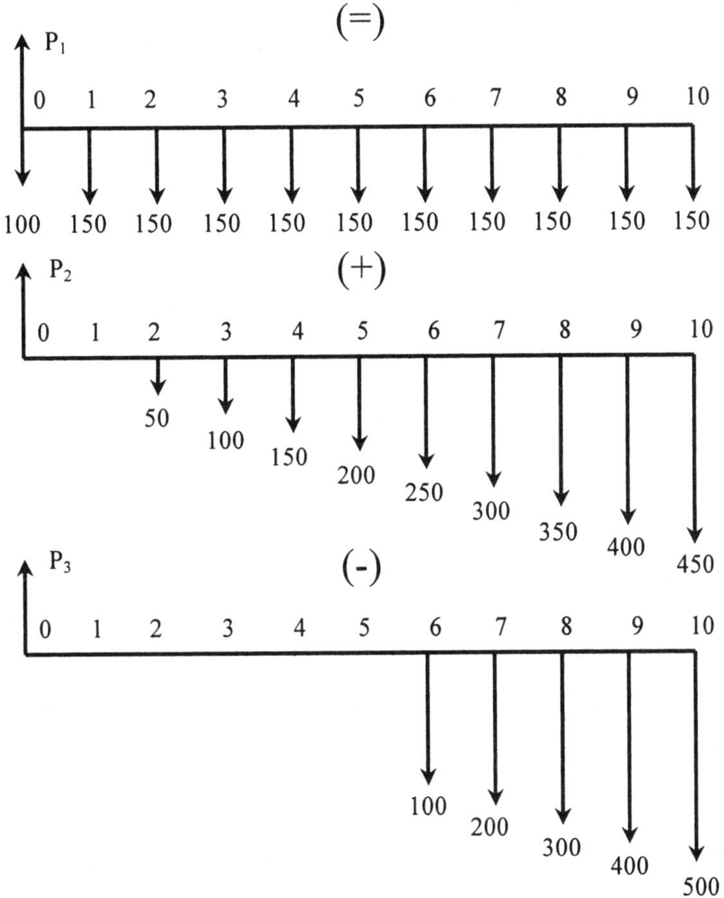

$P_1 = 100 + 150(P/A, 18\%, 10) = 774.10$
$P_2 = 50(P/G, 18\%, 10) = 717.60$
$P_3 = 100(P/G, 18\%, 6)(P/F, 18\%, 4) = 365.34$

$P = P_1 + P_2 + P_3 = \$1,126.36$

5-12

A couple wants to begin saving money for their child's education. They estimate that \$10,000 will be needed on the child's 18th birthday, \$12,000 on the 19th birthday, \$14,000 on the 20th birthday, and \$16,000 on the 21st birthday. Assume an 8% interest rate with only annual compounding. The couple is considering two methods of setting aside the needed money.

(a) How much money would have to be deposited into the account on the child's first birthday (note: a child's "first birthday" is celebrated one year after the child is born) to accumulate enough money to cover the estimated college expenses?

(b) What uniform annual amount would the couple have to deposit each year on the child's first through seventeenth birthdays to accumulate enough money to cover the estimated college expenses?

Solution

(a)

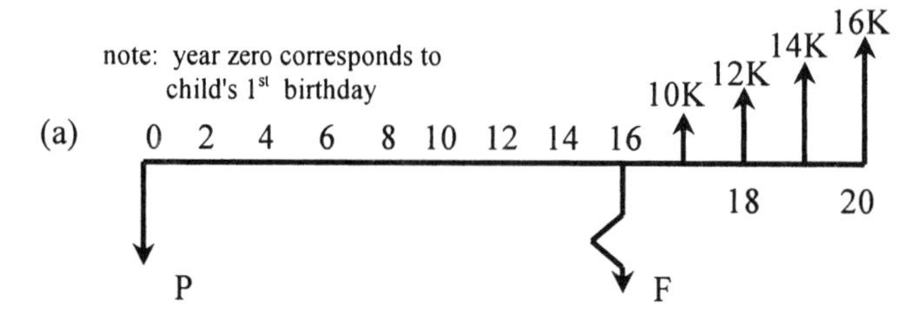

Let F = the \$'s needed at the beginning of year 16
= 10,000(P/A, 8%, 4) + 2,000(P/G, 8%, 4)
= 42,420

The amount needed today P = 42,420(P/F, 8%, 16) = \$12,382.40

(b)

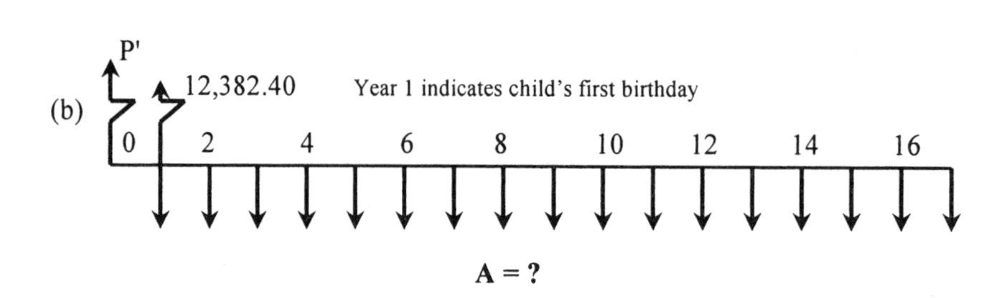

P' = 12,382.40(P/F, 8%, 1) = 11,464.86

A = 11,464.86(A/P, 8%, 17) = \$1,256.55

5-13

Assume you borrowed $50,000 at an interest rate of 1 percent per month, to be repaid in uniform monthly payments for 30 years. In the 163rd payment, how much of it would be interest, and how much of it would be principal?

Solution

In general, the interest paid on a loan at time t is determined by multiplying the effective interest rate times the outstanding principal just after the preceding payment at time t - 1.

To find the interest paid at time t = 163,(call it I_{163}) first find the outstanding principal at time t = 162 (call it P_{162}).

This can be done by computing the future worth at time t = 162 of the amount borrowed, minus the future worth of 162 payments. Alternately, compute the present worth, at time 162, of the 198 payments remaining.

The uniform payments are 50,000(A/P, 1%, 360) = $514.31, thus

P_{162} = 50,000(F/P, .01, 162) - 514.31(F/A, 1%, 162) = 514.31(P/A, 1%, 198) = $44,259.78

The interest is I_{163} = 0.01(44,259.78) = $442.59
and the principal in the payment is $514.31 - 442.59 = $71.72

5-14

A municipality is seeking a new tourist attraction, and the town council has voted to allocate $500,000 for the project. A survey shows that an interesting cave can be enlarged and developed for a contract price of $400,000. It would have an infinite life.

The estimated annual expenses of operation are:

Direct Labor	$30,000
Maintenance	15,000
Electricity	5,000

The price per ticket is to be based upon an average of 1000 visitors per month. If money is worth 8%, what should be the price of each ticket?

Solution

If the $100,000 cash, left over after developing the cave, is invested at 8%, it will yield a perpetual annual income of $8000. This $8000 can be used toward the $50,000 a year of expenses. The balance of the expenses can be raised through ticket sales, making the price per ticket

$42,000/12,000 tickets = $3.50/ticket

Alternate solution:

$$PW_{COST} = PW_{BENEFIT}$$
$$400{,}000 + (30{,}000 + 15{,}000 + 5{,}000)/.08 = 500{,}000 + T/.08$$
$$400{,}000 + 625{,}000 = 500{,}000 + T/.08$$
$$T = 525{,}000(.08)$$
$$= 42{,}000$$

Ticket Price = 42,000/ 12(1,000) = \$3.50

5-15

A middle-aged couple has made an agreement with Landscapes Forever Company, a gravesite landscaping and maintenance firm. The agreement states that Landscapes Forever will provide "deluxe landscaping and maintenance" for the couple's selected gravesite forever for an annual fee of \$1000. To arrange payment, the couple has set us a variable rate perpetual trust fund with their bank. The bank guarantees that the trust fund will earn a minimum of 5% per year. Assume that the services of Landscapes Forever will not be needed until after the wife has died, and that she lives to the ripe old age of 100.

(a) What is the smallest amount of money that the couple would have to deposit into the trust fund?

(b) Suppose that the couple made this minimum deposit on the wife's 50th birthday, and suppose that the interest rate paid by the trust fund fluctuated as follows:

Wife's Age	Interest Rate
50 - 54	5 %
55 - 64	10%
65 - 74	15%
75 - 84	20%

What is the largest sum of money that could be withdrawn from the trust fund on the wife's 85th birthday, and still have the perpetual payments to Landscapes Forever made?

Solution

(a) P = A/i = 1,000/.05 = $20,000

(b)

Age	i	Trust Fund Balance	
50-54	5%	20,000.00(F/P, 5%, 5) =	25,520.00
55-64	10%	25,520.00(F/P, 10%, 10) =	66,198.88
65-74	15%	66,198.88(F/P, 15%, 10) =	267,840.00
75-84	20%	267,840.00(F/P, 20%, 10) =	1,658,469.43

Therefore the largest sum which could be withdrawn from the trust fund is 1,658,469.43 - 20,000 = $1,632,469.43

5-16

A local car wash charges $3.00 per wash or the option of paying $12.98 for 5 washes, payable in advance with the first wash. If you normally washed your car once a month, would the option be worthwhile if your minimum attractive rate of return(MARR) is 12% compounded annually?

Solution

First, convert the effective annual MARR to its equivalent effective monthly rate:

$$(1.12)^{1/12} - 1 = 0.9489\%$$

Any measure of worth could now be used, but net present value is probably the easiest.

NPV = (-12.98 + 3.00) + 3.00(P/A, .9489%, 4) = $1.74 > 0

Therefore, the option is economical.

5-17

A project has a first cost of $14,000, uniform annual benefits of $2400, and a salvage value of $3000 at the end of its 10 year useful life. What is its net present worth at an interest rate of 12%?

Solution

PW = -14,000 + 2,400(P/A, 20%, 10) + 3,000(P/F, 20%, 10) = $526.00

5-18

A person borrows \$5,000 at an interest rate of 18%, compounded monthly. Monthly payments of \$180.76 are agreed upon.

(a) What is the length of the loan?
(Hint: it is an integral number of years.)

(b) What is the total amount that would be required at the end of the sixth month to payoff the entire loan balance?

Solution

(a)

$$P = A(P/A, i\%, n)$$
$$5{,}000 = 180.76(P/A, 1\tfrac{1}{2}\%, n)$$
$$(P/A, \tfrac{1}{2}\%, n) = 5{,}000/180.76$$
$$= 27.66$$

From the 1½% interest table n = 36 months = 6 years.

(b) $180.762 + 180.762(P/A, 1\tfrac{1}{2}\%, 30) = \$4{,}521.91$

5-19

A \$50,000 30-year loan with a nominal interest rate of 6% is to be repaid in payments of \$299.77 per month (for 360 months). The borrower wants to know how many payments, N*, he will have to make until he owes only half of the amount he borrowed initially. His minimum attractive rate of return (MARR) is a nominal 10% compounded monthly.

Solution

The MARR is irrelevant in this problem. The outstanding principal is always equal to the present worth of the remaining payments when the payments are discounted at the loan's effective interest rate.

Therefore, let N' be the remaining payments.

$$\tfrac{1}{2}(50{,}000) = 299.77(P/A, \tfrac{1}{2}\%, N)$$
$$(P/A, \tfrac{1}{2}\%, N) = 83.397$$
$$N = 108.30 \approx 108$$
$$\text{So, } N^* = 360 - N$$
$$= 252 \text{ payments}$$

5-20

A project has a first cost of \$10,000, net annual benefits of \$2000, and a salvage value of \$3000 at the end of its 10 year useful life. The project will be replaced identically at the end of 10 years, and again at the end of 20 years. What is the present worth of the entire 30 years of service if the interest rate is 10%?

Solution

PW of 10 years = - 10,000 + 2,000(P/A, 10%, 10) + 3,000(P/F, 10%, 10) = $3,445.76

PW of 30 years = 3,445.76[1 - (P/F, 10%, 30)] / [1 - (P/F, 10%, 10)] = $5,286.45

Alternate Solution:

PW of 30 years = [1 + (P/F, 10%, 10) +(P/F, 10%, 20)](-10,000) + 2,000(P/A, 10%, 30)
+ 3000 [(P/F, 10%, 10) +(P/F, 10%, 20) +(P/F, 10%, 30)]
= $5,286.45

5-21

The present worth of costs for a $5,000 investment with a complex cash flow diagram is $5265. What is the capitalized cost if the project has a useful life of 12 years, and the MARR is 18%?

Solution

Capitalized Cost = 5,265(A/P, 18%, 12)(P/A, 18%, ∞) = 5,265(.2086)(1/.18) = $6,102

5-22

A used car dealer tells you that if you put $1,500 down on a particular car your payments will be $190.93 per month for 4 years at a nominal interest rate of 18%. Assuming monthly compounding, what is the present price you are paying for the car?

Solution

A = 190.93 per period, i = .18/12 = .015, n = 4 x 12 = 48

P = 1,500 + 190.93(P/A, i%, 48)
= $8,000

5-23

What is the price of a 3-year Savings Certificate worth $5,000 three years hence, at 12 % interest, compounded continuously, with loss of interest if taken out before three years?

Solution

$P = Fe^{-rn} = \$5,000e^{-(0.12)3} = 5,000e^{-0.36} = \$3,488.50$

5-24

If the current interest rate on bonds of a certain type is 10% nominal, compounded semiannually, what should the market price of a $1,000 face value, 14 percent bond be? The bond will mature (pay face value) 6-1/2 years from today and the next interest payment to the bondholder will be due in 6 months.

Solution

Bi-yearly interest payment = .07(1,000) = $70

PV = $70(P/A, 5%, 13) + $1,000(P/F, 5%, 13) = $1,187.90

5-25

What is the Present Worth of a series that decreases uniformly, by $20 per year, from $400 in Year 11 to $220 in Year 20, if i equals 10 %?

Solution

PW = [400(P/A, 10%, 10) - 20(P/G, 10%, 10)](P/F, 10%, 10)
= $770.91

5-26

Many years ago BigBank loaned $12,000 to a local homeowner at a nominal interest rate of 4.5%, compounded monthly. The terms of the mortgage called for payments of $60.80 at the end of each month for 30 years. BigBank has just received the 300th payment, thus the loan has five more years to maturity. The outstanding balance is now $3,261.27.

Because BigBank currently charges a nominal 13% compounded monthly on home mortgages, it could earn a better return on its money if the homeowner would pay off the loan now; however, the bank realizes the homeowner has little economic incentive to do that with such a low interest rate on the loan. Therefore, BigBank plans to offer the homeowner a discount.

If the homeowner will pay today an amount of $3,261.27 - D, where D is the dollar amount of the discount, BigBank will consider the loan paid in full. If for BigBank the minimum attractive rate of return(MARR) is 10% (effective annual rate), what is the maximum discount, D, it should offer the homeowner?

Solution

The cash flows prior to now are irrelevant. The relevant cash flows are the following:

t	loan continues	paid off early	loan continues minus paid off early
0	0	+(3,261.27 - D)	-(3,261.27 - D)
1-6	+60.80		+60.80

Any measure of worth could be used.

The appropriate discount rate is the effective monthly MARR: $(1.1)^{1/12} - 1 = .00797$

Therefore, using NPV = 0 = -3261.27 + D + 60.80(P/A, .797%, 60)

D=$370.60

5-27
A resident will give money to his town to purchase a Vietnam veteran memorial statue and to maintain it at a cost of $500 per year forever. If an interest rate of 10% is used, and the resident gives a total of $15,000; how much can be paid for the statue?

Solution

Capitalized Cost = 15,000 = P + 500(P/A, 10%, ∞)
P = 15,000 - 500(1/.1) = $10,000

5-28
A rich widow decides on her 70th birthday to give most of her wealth to her family and worthy causes, retaining an amount in a trust fund sufficient to provide her with an annual end of year payment of $60,000. If she is earning a steady 10% rate of return on her investment, how much should she retain to provide these payments until she is 95(the last payment the day before she is 96)? If she dies on her 85th birthday, how much will remain in the trust fund?

Solution

P = 60K(P/A, 10%, 26) = 60K(9.161) = $549,660

P' = 60K(P/A, 10%, 11) = 60K(6.495) = $389,700

5-29

J.D. Homeowner has just bought a house with a 20-year, 9%, $70,000 mortgage on which he is paying $629.81 per month.

(a) If J.D. sells the house after ten years, how much must he give the bank to completely pay off the mortgage at the time of the 120th payment?
(b) How much of the first $379.33 payment on the loan is interest?

Solution

(a) P = 629.81 + 629.81(P/A, ¾%, 120) = $49,718.46

(b) $70,000 x 0.0075 = $525

5-30

Dolphin Inc. trains mine seeking dolphins in a 5-mine tank. They are considering purchasing a new tank. The U.S. Navy will pay $105,000 for each dolphin trained and a new tank costs $750,000 and realistic dummy mines cost $250,000. The new tank will allow the company to train 3 dolphins per year and will last 10 years costing $50,000 per year to maintain. Determine the net present value if the MARR equals 5%?

Solution

NPV = -Cost - Cost of Mines - Annual Maintenance(P/A, 5%, 10) + Income(P/A, 5%, 10)
= -750,000 - 250,000(5) - 50,000(P/A, 5%, 10) + 105,000(3)(P/A, 5%, 10)
= $46,330

5-31

Using an 8-year analysis and a 10% interest rate, determine which alternative should be selected, based on net present worth.

Alternative	A	B
First Cost	$5,300	$10,700
Uniform Annual Benefit	1,800	2,100
Useful life	4 years	8 years

Solution

NPW = PW(benefits) - PW(costs)

Alternative A:

NPW = 1,800(P/A, 10%, 8) - 5,300 - 5,300(P/F, 10%, 4)
= $683.10

Alternative B:

NPW = 2,100(P/A, 10%, 8) - 10,700 = $503.50

Select alternative A

5-32
Three purchase plans are available for a new car.

Plan A: $5,000 cash immediately
Plan B: $1,500 down and 36 monthly payments of $116.25
Plan C: $1,000 down and 48 monthly payments of $120.50

If a customer expects to keep the car five years and her minimum attractive rate of return (MARR) is 18% compounded monthly, which payment plan should she choose?

Solution

Note that in all cases the car is kept 5 years that is the common analysis period.

$i = 18\%/12 = 1\frac{1}{2}\%$

$PWC_A = \$5{,}000$

$PWC_B = 1{,}500 + 116.25(P/A, 1\frac{1}{2}\%, 36) = \$4{,}715{,}59$

$PWC_C = 1{,}000 + 120.50(P/A, 1\frac{1}{2}\%, 48) = \$5{,}102.18$

Therefore Plan B is best

5-33
Given the following three mutually exclusive alternatives.

	Alternative		
	A	B	C
Initial Cost	$50	$30	$40
Annual Benefits	15	10	12
Useful Life(years)	5	5	5

What alternative is preferable if i = 10%?

Solution

	A	B	C
Initial Cost	50.00	30.00	40.00
Annual Benefits	15.00	10.00	12.00
Useful Life(years)	5	5	5
Present Worth Benefits	56.87	37.91	45.49
Present Worth Costs	50.00	30.00	40.00
Net Present Worth = PWB - PWC	6.87	7.91	5.49

Choose C

5-34

Consider two investments:

(1) Invest $1000 and receive $110 at the end of each month for the next 10 months.
(2) Invest $1200 and receive $130 at the end of each month for the next 10 months.

If this were your money, and you want to earn at least 12% interest on your money, which investment would you make, if any? What nominal interest rate do you earn on the investment you choose? Solve by present worth analysis.

Solution

i = 12%/12 = 1% per month

Alternative 1: NPW = 110(P/A, 1%, 10) - 1,000 = $41.81
Alternative 2: NPW = 130(P/A, 1%, 10) - 1,200 = $31.23

Choose Alternative 1 → Maximum NPW

Nominal Interest: NPW = 0 = -1,000 + 110(P/A, i%, 10)
(P/A, i%, 10) = 9.1

From tables: i ≅ 1.75%

Nominal interest = 1.75% x 12 mo. = 21%

5-35

A farmer has just purchased a tractor for which he had to borrow $20,000. The bank has offered the following choice of payment plans determined using an interest rate of 8%. If the farmer's minimum attractive rate of return (MARR) is 15%, which plan should he choose?

Plan A: $5,010 per year for 5 years
Plan B: $2,956 per year for 4 years plus $15,000 at end of 5 years
Plan C: Nothing for 2 years, then $9048 per year for 3 years

Solution

$PWC_A = 5{,}010(P/A, 15\%, 5) = \$16{,}794$

$PWC_B = 2{,}956(P/A, 15\%, 4) + 15{,}000(P/F, 15\%, 5) = \$15{,}897.$

$PWC_C = 9{,}048(P/A, 15\%, 3)(P/F, 15\%, 2) = \$15{,}618$

Plan C is lowest cost plan

5-36

Projects A and B have first costs of $5,000 and $9,000, respectively. Project A has net annual benefits of $2,500 during each year of its 5 year useful life, after which it can be replaced identically.

Project B has net annual benefits of $3,300 during each year of its 10 year life. Use present worth analysis, an interest rate of 30% per year, and a 10 year analysis period to determine which project to select.

Solution

$PW_A = -5{,}000[1 + (P/F, 30\%, 5)] + 2{,}500(P/A, 30\%, 10) = \$1{,}382.20$

$PW_B = -9{,}000 + 3{,}300(P/A, 30\%, 10) = \$\ 1{,}202.08$

Select A because of higher present worth of benefits.

5-37

The lining of a chemical tank in a certain manufacturing operation is replaced every 5 years at a cost of $5,000. A new type lining is now available which would last 10 years, but costs $13,000. The tank needs a new lining now and you intend to use the tank for 40 years, replacing linings when necessary. Compute the present worth of costs of 40 years of service for the 5-year and 10-year linings if i = 10%.

Solution

PW 5 yr Lining:

PW = [5,000(A/P, 10%, 5)](P/A, 10%, 40) = $12,898.50

PW 10 yr Lining:

PW = [13,000(A/P, 10%, 10)](PA, 10%, 40) = $20,683.50

5-38

A manufacturing firm has a before-tax minimum attractive rate of return (MARR) of 12% on new investments. What uniform annual benefit would Investment B have to generate to make it preferable to Investment A?

Year	Investment A	Investment B
0	- $60,000	- $45,000
1 - 6	+15,000	?

Solution

NPW of A = - 60 + 15(P/A, 12%, 6) = 1.665

NPW of B $\geq$ 1.665 = - 45 + B(P/A, 12%, 6)
$\therefore$ B = 11,351

B > $11,351 per year

5-39

The city council wants the municipal engineer to evaluate three alternatives for supplementing the city water supply. The first alternative is to continue deep well pumping at an annual cost of $10,500. The second alternative is to install an 18" pipeline from a surface reservoir. First cost is $25,000 and annual pumping cost is $7000.

The third alternative is to install a 24" pipeline from the reservoir at a first cost of $34,000 and annual pumping cost of $5000. Life of all alternatives is 20 years. For the second and third alternatives, salvage value is 10% of first cost. With interest at 8%, which alternative should the engineer recommend? Use present worth analysis.

Solution

Fixed output, therefore minimize cost.

Year	DEEPWELL	18" PIPELINE	24" PIPELINE
0		-25,000	-34,000
1-20	-10,500	-7,000	-5,000
20		+2,500	+3,400

Deepwell: PWC = - 10,500(P/A, 8%, 20) -$103,089

18" Pipeline: PW of Cost = -25,000 - 7,000(P/A, 8%, 20) + 2,500(P/F, 8%, 20) = -$493,190

24" Pipeline: PW of Cost = -34,000 - 5,000(P/A, 8%, 20) + 3,400(P/F, 8%, 20) = -$82,361

Choose 24" Pipeline

5-40

A magazine subscription is $12 annually, or $28 for a 3-year subscription. If the value of money is 12%, which choice is best?

Solution

28 <?=?> 12 + 12(P/A, 12%, 2) = 12 + 12(1.69) = 32.28

Choose 3 yr subscription because 28 < 32.28

5-41

Two alternatives are being considered for recovering aluminum from garbage. The first has a capital cost of $100,000, a first year maintenance cost of $5000, with maintenance increasing by $1500 per year for each year after the first.

The second has a capital cost of $120,000, a first year maintenance cost of $3000, with maintenance increasing by $1000 per year after the first.

Revenues from the sale of aluminum are $20,000 in the first year, increasing $2000 per year for each year after the first. Life of both alternatives is 10 years. There is no salvage value. The before-tax Minimum Attractive Rate of Return is 10%. Using Present Worth Analysis determine which alternative is preferred.

Solution

Alternative 1: NPW = -100,000 +(20,000 - 5,000)(P/A, 10%, 10) = \$3,620.50

Alternative 2: NPW = -120,000 +(20,000 - 3,000)(P/A, 10%, 10) = \$7,356.00

Choose Alternative 2 → Maximum. NPW

5-42

A brewing company is deciding between two used filling machines as a temporary measure, before a plant expansion is approved and completed. The two machines are:

(a) The Kram Filler. Its initial cost is \$85,000, and the estimated annual maintenance is \$8000.

(b) The Zanni Filler. The purchase price is \$42,000, with annual maintenance costs of \$8000.

The Kram filler has a higher efficiency, compared with the Zanni, and it is expected that the savings will amount to \$4000 per year if the Kram filler is installed. It is anticipated that the filler machine will not be needed after 5 years, and at that time, the salvage value for the Kram filler would be \$25,000, while the Zanni would have little or no value.

Assuming a minimum attractive rate of return (MARR) of 10%, which filling machine should be purchased?

Solution

Fixed output, therefore minimize costs

Kram:

NPW = 25,000(P/F, 10%, 5) - 85,000 - 4,000(P/A, 10%, 5)
= -\$84,641.5 (or a PWC \$84,641.50)

Zani:

NPW = -42,000 - 8,000(P/A, 10%, 5)
= -\$72,328 (or a PWC of \$72,328)

Therefore choose the Zani filler.

5-43
Two technologies are currently available for the manufacture of an important and expensive food and drug additive. The two can be described as follows:

Laboratory A.
Is willing to release the exclusive right to manufacture the additive in this country for \$50,000 payable immediately, and a \$40,000 payment each year for the next 10 years.
The production costs are \$1.23 per unit of product.

Laboratory B.
This laboratory is also willing to release similar manufacturing rights. They are asking for the following schedule of payments:
On the closing of the contract, \$10,000.
From years 1 to 5, at the end of each year, a payment of \$25,000 each.
From years 6 to 10, also at the end of each year, a payment of \$20,000.
The production costs are \$1.37 per unit of product.

Neither lab is to receive any money after 10 years for this contract. It is anticipated there will be an annual production of 100,000 items for the next 10 years. On the basis of analyses and trials, the products of A and B are practically identical in quality. Assuming a MARR of 12%, which lab should be chosen?

Solution

Laboratory A: The annual production cost = 1.23 x 100K = \$123K

PWC = 50,000 + [40,000 + 123,000](P/A, 12%, 10) = \$970,950

Laboratory B: The annual production cost = 1.37 x 100K = \$137K

PWC = 10,000 + [25,000 + 137,000](P/A, 12%, 5)
+ [20,000 + 137,000](P/A, 12%, 5)(P/F, 12%, 5) = \$915,150

Therefore choose Laboratory B.

5-44
An engineering analysis by net present worth (NPW) is to be made for the purchase of two devices A and B. If an 8% interest rate is used, recommend the device to be purchased.

	Cost	Uniform Annual Benefit	Salvage	Useful Life
Device A	\$600	\$100	\$250	5 years
Device B	700	100	180	10 years

Solution

Device A:

NPW = 100(P/A, 8%, 10) + 250((P/F, 8%, 10) - 600 - [600 - 250](P/F, 8%, 5) = -$51.41

Device B:

NPW = 100(P/A, 8%, 10) + 180(P/F, 8%, 10) - 700 = $54.38

Select device B

5-45

A company decides it must provide repair service for the equipment it sells. Based on the following, which alternative for providing repair service should be selected?

Alternative	Net Present Worth
A	-$15,725
B	-6,657
C	-8,945

Solution

None of the alternatives look desirable, but since one of the alternatives must be chosen (the do nothing alternative is not available), choose the one that maximizes NPW. Thus the best of the three alternatives is B.

5-46

A firm is considering the purchase of a new machine to increase the output of an existing production process. Of all the machines considered, the management has narrowed the field to the machines represented by the cash flows shown as follows:

Machine	Initial Investment	Annual Operating Cost
5	$50,000	$22,500
2	60,000	20,540
4	75,000	17,082
1	80,000	15,425
3	100,000	11,374

If each of these machines provides the same service for 3 years and the minimum attractive rate of return is 12%, which machine should be selected?

Solution

Minimize the PW of Cost:

Machine	Initial Investment	Operating Costs x (P/A, 12%, 3)	PW of Costs
5	-50,000	+ 22,500 (2.402)	= 104,045
2	-60,000	+ 20,540(2.402)	= 109,337
4	-75,000	+ 17,082(2.402)	= 116,031
1	-80,000	+ 15,425(2.402)	= 117,051
3	-100,000	+ 11,374(2.402)	= 127,320

Select machine 5

Chapter 6

Annual Cash Flow Analysis

6-1

While in college Ellen received $10,000 in student loans at 5% interest. She will graduate in June and is expected to begin repaying the loans in either 5 or 10 equal annual payments. Compute her yearly payments for both repayment plans.

Solution

5 YEARS	10 YEARS
A = P(A/P, i, n)	A = P(A/P, i, n)
= 10,000(A/P, 5%, 5)	= 10,000(A/P, 5%, 10)
= $2,310.00	= $1,295.00

6-2

Suppose you wanted to buy a $100,000 house. You have $20,000 cash to use as the down payment. The bank offers to loan you the remainder at 6% nominal interest. The term of the loan is 20 years. Compute your monthly loan payment assuming the payment is the same for all months.

Solution

Amount of loan: $100,000 - $20,000 = $80,000

i = 6%/12 = ½% per month   n = 20 x 12 = 240 periods

A = 80,000(A/P, ½%, 240)
= $572.80 per month

6-3

Lester Peabody decides to install a fuel storage system for his farm that will save him an estimated 6.5 cents/gallon on his fuel cost. He uses an estimated 20,000 gallons/year on his farm. Initial cost of the system is $10,000 and the annual maintenance is a uniform gradient amount of $25. After a period of 10 years the estimated salvage is $3,000. If money is worth 12%, is it a wise investment?

Solution

EUAC = (10,000 - 3,000)(A/P, 12%, 10) + 3,000(.12) + 25(A/G, 12%, 10)
= $1,688.63

EUAB = 20,000(.065) = $1,300

EUAW = -$388.63 ∴ not a wise investment

6-4

The returns for a business for five years are as follows: $8,250, $12,600, $9,750, $11,400, and $14,500. If the value of money is 12%, what is the equivalent uniform annual benefit for the five-year period?

Solution

PW = 8,250(P/F, 12%, 1) + 12,600(P/F, 12%, 2) + 9,750(P/F, 12%, 3)
+ 11,400(P/F, 12%, 4) + 14,500(P/F, 12%, 5)
= $39,823

EUAB = 39,823(A/P, 12%, 5) = $11,047

6-5

The local loan shark has loaned you $1000. The interest rate you must pay is 20%, compounded monthly. The loan will be repaid by making 24 equal monthly payments. What is the amount of each monthly payment?

Solution

A = 1,000(A/P, 20%/12, 24)

There is no 1-2/3% compound interest table readily available. Therefore the capital recovery factor must be calculated.

$(A/P, 1.666\%, 24) = [0.01666(1.01666)^{24}] / [(1.01666)^{24} - 1] = 0.050892$

A = 1,000(0.050892) = $50.90

6-7

Several companies offer "instant cash" plans to holders of their credit cards. A typical plan permits card holders to "draw" cash up to a preset limit. At the time the cash is drawn, a special charge of 4% of the amount drawn is charged to the card holder's account. Then the card holder repays the debt (the original amount drawn plus the special charge) by making a series of equal monthly payments. Each month the company adds a finance charge of 1½% of the previous month's unpaid balance to the account balance. If the card holder "draws" $150, a $6 special charge will be made and the card holder will make a series of monthly payments of $9.95.

(a) How many payments will be required?
(b) What "true" (effective) annual interest rate does the card holder pay?

Solution

(a) 9.95(P/A, 1½%, n) = 156
(P/A, 1½%, n) = 15.678

From compound interest tables n = 18 + a very slight amount
PW of payments = 9.95(P/A, 1½%, 18) = $155.95 for 18 payments
FW of balance = 0.05(F/P, 1½%, 19) = $0.07
So there are 18 payments of $9.95 and a final payment of 7 cents.

(b) 150 = 9.95(P/A, i%, 18) + 0.07(P/F, i%, 19)

solve for i, solution using tables:
try i = 1¾% NPW = -150 + 9.95(P/A 1¾%, 18) + 0.07(P/F, 1¾%, 19) = $2.55

try i = 2% NPW = -150 + 9.95(P/A, 2%, 18) + 0.07(P/F, 2%, 19) = -$0.78

interpolate: i = 1¾% + [(2.55 - 0)/(2.55 – (-0.78)](¼%) = 1.9414%

Effective annual interest rate = $(1 + 0.019414)^{12}$ - 1 = 0.2595 = 25.95 %

6-8

If $15,000 is deposited into a savings account that pays 4% interest compounded quarterly, how much can be withdrawn each quarter for five years?

Solution

i = 4%/4 = 1% per quarter n = 5 x 4 = 20 periods

A = 15,000(A/P, 1%, 20)
= $831.00 per quarter

6-9

What uniform annual payment for 12 years is equivalent to receiving all of these:

\$ 3,000	at the end of each year for 12 years
20,000	today
4,000	at the end of 6 years
800	at the end of each year forever
10,000	at the end of 15 years

Use an 8% interest rate.

Solution

A_1 = \$3,000
A_2 = 20,000(A/P, 8%, 12) = \$2,654
A_3 = 4,000(P/F, 8%, 6)(A/P, 8%, 12) = \$334.51
A_4 =(800/.08)(A/P, 8%, 12) = \$1,327
A_5 = 10,000(P/F, 8%, 15)(A/P, 8%, 12) = \$418.27

$$\sum_{i=1}^{n} Ai = 3{,}000 + 2{,}654 + 334.51 + 1{,}327 + 418.27 = \$7{,}733.78$$

6-10

For the following cash flow diagram, which equation properly calculates the uniform equivalent?

(a) A = 100(A/P, i, 3) + 100(A/F, i, 3)
(b) A = 100(A/P, i, 15)
(c) A = 100(A/F, i, 3) + 100(A/P, i, 15)
(d) A = 100(A/F, i, 3) + 100(A/F, i, 15)
(e) A = 100(A/F, i, 3)

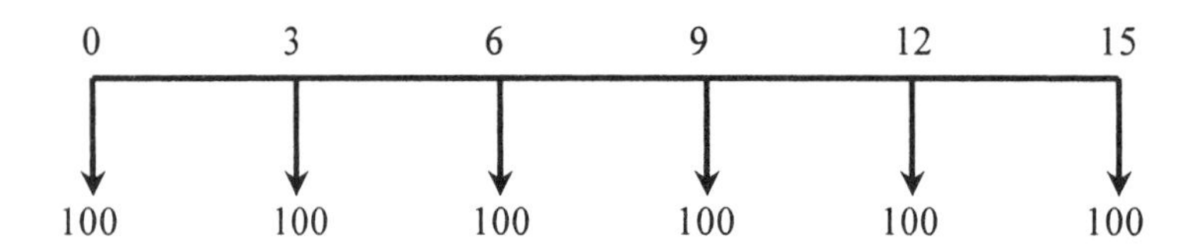

Solution

The correct equation is (c).

6-11

A project has a first cost of \$75,000, operating and maintenance costs of \$10,000 during each year of its 8 year life, and a \$15,000 salvage value. What is its equivalent uniform annual cost (EUAC) if the interest rate is 12%?

Solution

EUAC = 75,000(A/P, 12%, 8) + 10,000 - 15,000(A/F, 12%, 8) = \$23,878.00

6-12

A land surveyor just starting in private practice needs a van to carry crew and equipment. He can lease a used van for \$3,000 per year, paid at the beginning of each year, in which case maintenance is provided. Alternatively, he can buy a used van for \$7,000 and pay for maintenance himself. He expects to keep the van three years at which time he could sell it for \$1,500. What is the most he should pay for uniform annual maintenance to make it worthwhile buying the van instead of leasing it, if his MARR is 20%?

Solution

Lease:

EUAC = 3,000(F/P, 20%, 1) = 3,000(1.20) = 3,600

Buy:

EUAC = 7,000(A/P, 20%, 3) + M - 1,500(A/F, 20%, 3)
M = 3,600 - 2,910.85
= \$ 689.15

6-13

A foundation supports an annual seminar on campus by using the earnings of a \$50,000 gift. It is felt that 10% interest will be realized for 10 years, but that plans should be made to anticipate an interest rate of 6% after that time. What uniform annual payment may be established from the beginning, to fund the seminar at the same level into infinity?

Solution

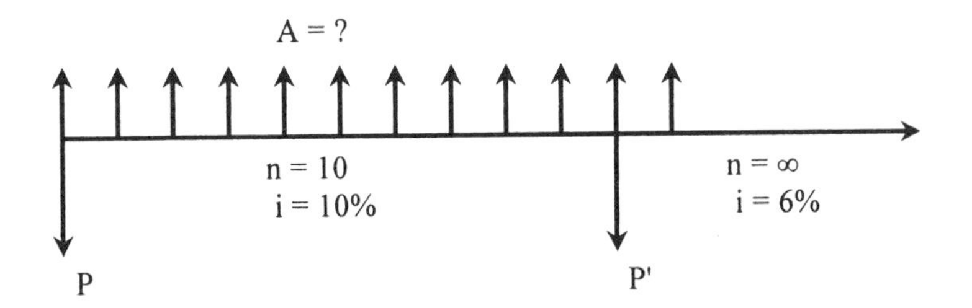

Assume first seminar occurs at time of deposit.

$$P' = A/i = A/.06$$

$$P = A + A(P/A, 10\%, 10) + P'(P/F, 10\%, 10)$$
$$50{,}000 = A + 6.145A + (A/.06) \times .3855$$
$$13.57A = 50{,}000$$

$$A = \$3{,}684.60$$

6-14

A consumer purchased new furniture by borrowing \$1,500 using the store's credit plan which charges 18% compounded monthly.

(a) What are the monthly payments if the loan is to be repaid in 3 years?
(b) How much of the first payment is interest?
(c) How much does the consumer still owe just after making the 20th payment?

Solution

(a) $i = 18\%/12 = 1\frac{1}{2}\%$ per month, $n = 12 \times 3 = 36$ periods

$$A = 1{,}500(A/P, 1\tfrac{1}{2}\%, 36)$$
$$= \$54.30 \text{ per month}$$

(b) Int. = Principal x Monthly interest rate
Int. = 1,500 x 0.015
= \$22.50

(c) $P = 54.30(P/A, 1\tfrac{1}{2}\%, 16)$
$= \$767.31$

6-15

A 30-year mortgage of \$100,000 at a 6% interest rate had the first payment made on September 1, 1999. What amount of interest was paid for the 12 monthly payments of 2002?

Solution

Monthly payment $A = 100{,}000(A/P, \tfrac{1}{2}\%, 360) = \$599.55$

Interest periods remaining Jan 1, 2002 = 331
Jan 1, 2003 = 319

P' = 599.55(P/A, ½%, 331) = 599.55(161.624) = 96,901.67

P" = 599.55(P/A, ½%, 319) = 599.55(159.257) = 95,482.53

Interest = 599.55(12) - (96,901.67 – 95,482.53) = $5,775.46

6-16

A grateful college graduate makes a donation of $2,000 now and will pay $37.50 per month for 10 years to establish a scholarship. If interest in the fund is computed at 9%, what annual scholarship may be established? Assume the first scholarship will be paid at the end of the first year.

Solution

P = 2,000 + 37.50(P/A, ¾%, 120)
= 4,960.33
A = Pi
= 4,960.33(.09)
= $446.43 scholarship

6-17

A rich folk singer has donated $500,000 to endow a university professorial chair in Bohemian Studies. If the money is invested at 8.5%, how much can be withdrawn each year, ad infinitum, to pay the Professor of B.S.?

Solution

$A = 500{,}000(A/P, 8.5\%, \infty) = 500{,}000(.085) = \$42{,}500$

6-18

For the cash flow shown below, what would an equivalent sum of money be
(a) Now?
(b) Two years from now?
(c) As a five year annuity, starting at the end of the first year?

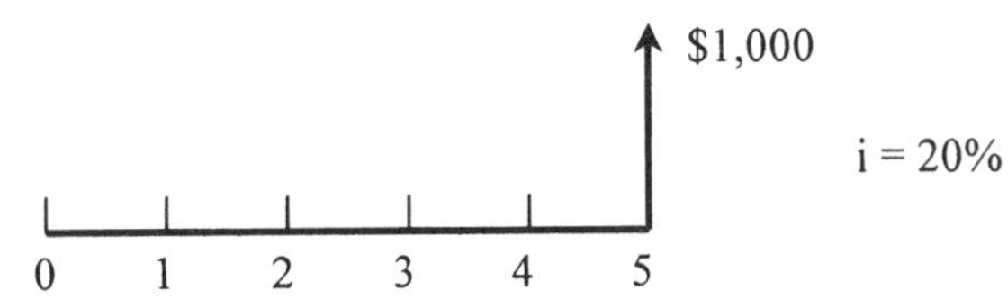

Solution

(a) PV = $1,000(P/F, 20%, 5) = $401.90
(b) $401.90(F/P, 20%, 2) = $578.74
(c) A = $1,000(A/F, 20%, 5) = $134.40

6-19

A project requires an initial investment of $10,000 and returns benefits of $6,000 at the end of every 5th year thereafter. If the minimum attractive rate of return (MARR) is 10%, should the project be undertaken? Show supporting calculations.

Solution

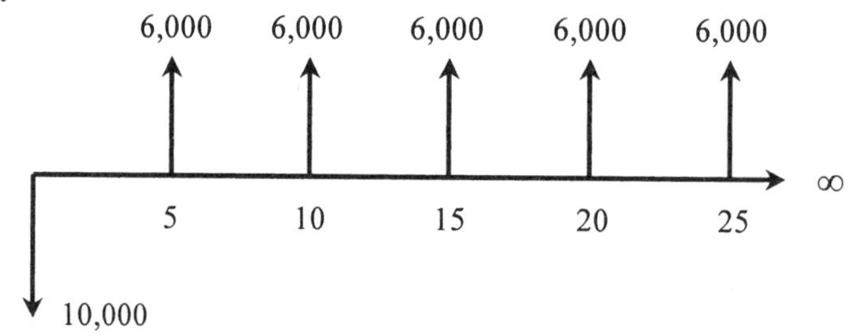

$$\text{EUAW} = 6{,}000(A/F, 10\%, 5) - 10{,}000(A/P, 10\%, \infty)$$
$$= -\$17.20$$

The project should not be undertaken.

6-20

On January 1st an engineering student projects a need of $2,400 on December 31st. What amount must be deposited in the credit union each month if the interest paid is 3%, compounded monthly?

Solution

$$A = 2{,}400(A/F, \tfrac{1}{4}, 12)$$
$$= \$197.28$$

6-21

Given:

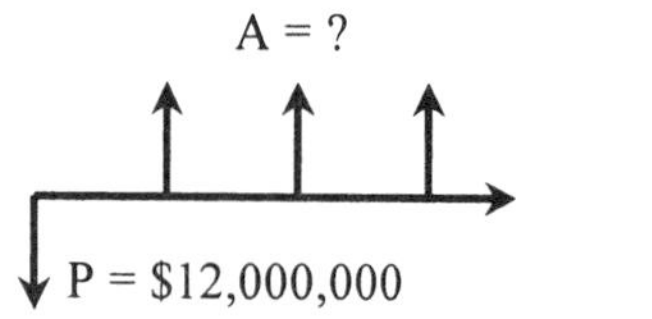

$n = \infty$
$i = 3\%$

Find: A

Solution

$$A = Pi$$
$$= 12{,}000{,}000(0.03)$$
$$= \$360{,}000$$

6-22
Assuming monthly payments, which would be the better financing plan on the same \$9,000 car?

(a) 6% interest on the full amount for 48 months, compounded monthly.
(b) A \$1,000 rebate(discount) and 12% interest on the remaining amount for 48 months, compounded monthly.

Solution

(a) A = 9,000(A/P, ½%, 48) = \$211.50/mo.

(b) A = 8,000(A/P, 1%, 48) = \$210.40/mo.

Choose plan b.

6-23
The initial cost of a pickup truck is \$11,500 and will have a salvage value of \$4,000 after five years. Maintenance is estimated to be a uniform gradient amount of \$150 per year (first year maintenance = zero), and the operation cost is estimated to be 30 cents per mile for 300 miles per month. If money is worth 12%, what is the equivalent uniform annual cost (EUAC) for the truck expressed as a monthly cost?

Solution

EUAC = [(11,500 - 4,000)(A/P, 1%, 60) + 4,000(.01) + 150(A/G, 12%, 5)]/12 + 300(.30)
= \$318.69 per month

6-24
Twenty five thousand dollars is deposited in a bank trust account that pays 18% interest, compounded semiannually. Equal annual withdrawals are to be made from the account, beginning one year from now and continuing forever. Calculate the maximum amount of the equal annual withdrawal.

Solution

i =18%/2 = 9% per semi-annual period

A = Pi = 25,000(.09) = 2,250

W = 2,250(F/A, 9%, 2) = \$4,702.50

6-25

A truck, whose price is \$18,600, is being paid for in 36 uniform monthly installments, including interest at 10 percent. After making 13 payments, the owner decides to pay off the remaining balance of the purchase price in one lump sum. How much is the lump sum?

Solution

In problems like this the lump sum is the present worth of all the future(unpaid) payments. So to solve the problem compute the payment and then compute the PW of the unpaid payments at the stated interest rate.

A = 18,600(A/P, .83%, 36)
= $18{,}600[(.00833(1 + .00833)^{36})/((1 + .00833)^{36} - 1)]$
= \$600.22

After 13 months: 36 - 13 = 23 payments remain

P = 600.22(P/A, .83%, 23)
= $600.22[((1 + .00833)^{23} - 1)/(.00833(1 + .00833)^{23})]$
= \$12,515.45

6-26

If the interest rate is 10% and compounding is semiannual, what series of equal annual transactions is equivalent to the following series of semiannual transactions? The first of the equal annual transactions is to occur at the end of the second year and the last at the end of the fourth year.

Time (yr)	0		1		2		3		4		5	5½
Period	0	1	2	3	4	5	6	7	8	9	10	11
Cash Flow	\$0	600	500	400	300	200	100	300	500	700	900	1100

Solution

P = 600(P/A, 5%, 5) - 100(P/G, 5%, 5) + [100(P/A, 5%, 6) + 200(P/G, 5%, 6)](P/F, 5%, 5)
= 4,046.80

Effective i $=(1 + 0.10/2)^2 - 1 = 10.25\%$

Sum at end of Year 1: F = P(F/P, 10.25%, 1) = 4,461.60

Equal Annual Payments: A = 4,461.60(A/P, 10.25%, 3) = \$1,802.04

6-27

A tractor costs $12,500 and will be used for five years when it is estimated a salvage value of $4,000 will be appropriate. Maintenance cost is estimated to be a $100 the first year and increase by $100 each year thereafter. If a 12% interest rate is used, what is the equivalent uniform annual cost (EAC) for the tractor?

Solution

$$\begin{aligned} \text{EUAC} &= 12{,}500(A/P, 12\%, 5) + 100 + 100(A/G, 12\%, 5) - 4{,}000(A/F, 12\%, 5) \\ &= \$3{,}115.40 \end{aligned}$$

6-28

If Zoe won $250,000 the last week in February, 1996 and invested it by March 1, 1996 in a "sure thing" which pays 8% interest, compounded annually, what uniform annual amount can he withdraw on the first of March for 15 years starting in 2004?

Solution

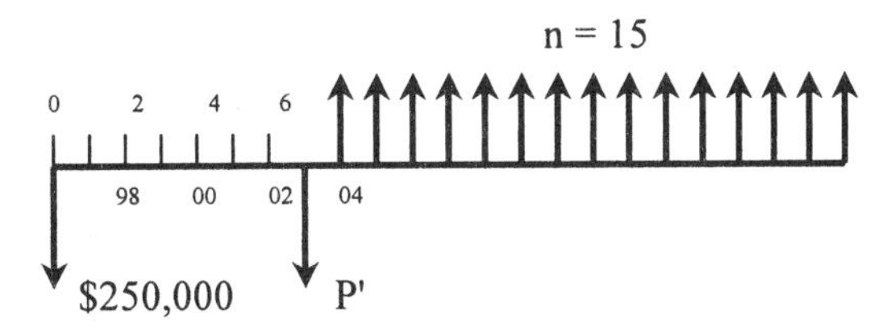

$$P' = 250K(F/P, 8\%, 7) = \$428{,}500$$
$$A = 250K(F/P, 8\%, 7)(A/P, 8\%, 15) = \$50{,}048.80$$

6-29

A machine, with a first cost of $20,000, is expected to save $1,500 in the first year of operation and the savings should increase by $200 every year until (and including) the ninth year, thereafter the savings will decrease by $150 until (and including) the 16th year.

Using equivalent uniform annual worth, is this machine economical? Assume a minimum attractive rate of return of 10%.

Solution

There are a number of equations that could be written.
Here's one:

$$\begin{aligned} \text{EuAW} &= -20{,}000(A/P, 10\%, 16) + [1{,}500(P/A, 10\%,9) + 200(P/G, 10\%, 9)](A/P, 10\%, 16) \\ &\quad + [1{,}450(P/A, 10\%, 7) - 150(P/G, 10\%, 7)](P/F, 10\%, 9)(A/P, 10\%, 16) \\ &= -\$676.79\text{, the machine is not economical} \end{aligned}$$

6-30

Calculate the equivalent uniform annual cost of the following schedule of payments.

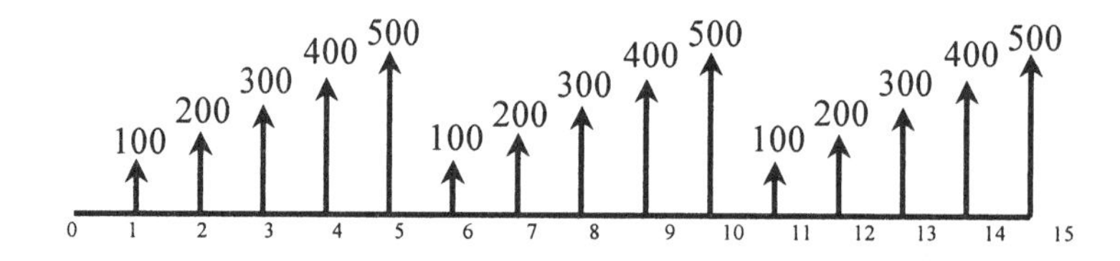

Solution

Since payments repeat every five years, analyze for 5 years only.

A = 100 + 100(A/G, 8%, 5) = $284.60

6-31

A college is willed $100,000 to establish a permanent scholarship. If funds are invested at 6% and all funds earned are disbursed yearly, what will be the value of the scholarship in the 6th year of operation?

Solution

A = Pi = 100,000(0.06) = $6,000 for any year

6-32

The UNIFORM EQUIVALENT of the cash flow diagram shown is given by which one of the following five answers?

(a) 50(A/G, i, 8)
(b) 50(A/G, i, 9)
(c) 50(A/G, i, 10)
(d) 50(A/G, i, 9)(F/A, i, 9)(A/F, i, 10)
(e) 50(P/G, i, 8)(P/F, i, 1)(A/P, i, 10)

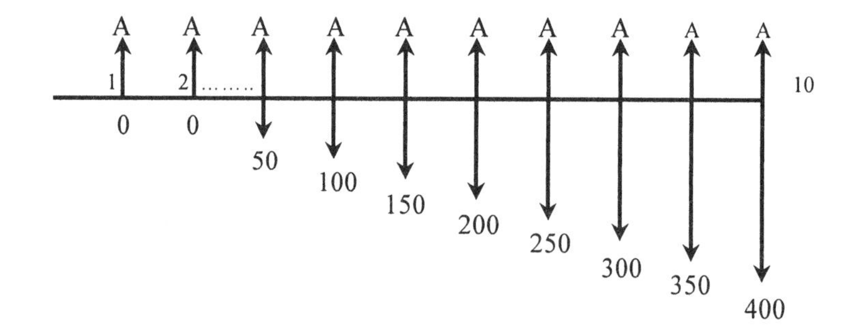

Solution

Note these two concepts:

1) The G series is 9 periods long
2) The uniform equivalent is 10 periods long

The answer is(d)

6-33

To get started, Econ Engineering has just borrowed $500,000 that will be paid off in 20 end-of-quarter payments. If interest is 18% compounded monthly, what will be the size of each loan payment?

Solution

$i = 18\%/12 = 1\frac{1}{2}\%$ $\quad\quad$ $n = 60$ monthly interest periods

$$P = 500{,}000 = X(A/F, 1\tfrac{1}{2}\%, 3)(P/A, 1\tfrac{1}{2}\%, 60)$$
$$= X(.3284)(39.380)$$
$$X = \$38{,}663$$

6-34

The cost of an automobile is $9,000 and after a period of three years it will have an estimated salvage value of $5,200. A downpayment of $1,000 will be used to purchase the car. It is desired to make the monthly payments, at 12% interest, a value such as to reduce the unpaid balance to exactly the amount of the salvage value after three years. What is the amount of the monthly payment?

Solution

$$\text{PW of Salvage} = 5{,}200(P/F, 1\%, 36) = \$3{,}634.28$$

$$P = 9{,}000 - 3{,}634.28 = \$5{,}365.72$$

$$A = 5{,}365.72(A/P, 1\%, 36)$$
$$= \$178.14 \text{ monthly payment}$$

6-35

Joyce and Bill purchased a four unit apartment house and as part of the financing obtained a $100,000 loan at 9% nominal annual interest, with equal monthly payments for 20 years.

What is their monthly payment?

Solution

$i = 9\%/12 = \frac{3}{4}\%$ per month $\qquad n = 20 \times 12 = 240$ periods

$A = 100{,}000(A/P, \frac{3}{4}\%, 240)$
$= \$900$ per month

6-36

The initial cost of a van is $12,800 and will have a salvage value of $5,500 after five years. Maintenance is estimated to be a uniform gradient amount of $120 per year (with no maintenance costs the first year), and the operation cost is estimated to be 36 cents/mile for 400 miles/month. If money is worth 12%, what is the equivalent uniform annual cost (EUAC) for the van, expressed as a monthly cost?

Solution

$EUAC = (12{,}800 - 5{,}500)(A/P, 12\%, 5) + (5{,}500)(.12) + 120(A/G, 12\%, 5) + .36(400)(12)$
$= 4{,}626/12$
$= \$385.50/\text{month}$

6-37

An engineering student purchased a 2-year-old car that sold new for $8,000. The car depreciated 25% per year. The student made a downpayment of $1000 and obtained a 36 month loan at 15% interest, compounded monthly. What were the monthly payments?

Solution

year 1 depreciation = (.25)(8,000) = 2,000 $\qquad$ 8,000 - 2,000 = $6,000
year 2 depreciation = (.25)(6,000) = 1,500 $\qquad$ 6,000 - 1,500 = $4,500 sale price

if downpayment = 1,000 ; loan = 3,500

$i = 15\%/12 = 1\frac{1}{4}\%$ $\qquad n = 12 \times 3 = 36$ periods

$A = \$3{,}500(A/P, 1\frac{1}{4}\%, 36)$
$= \$121.45$

6-38

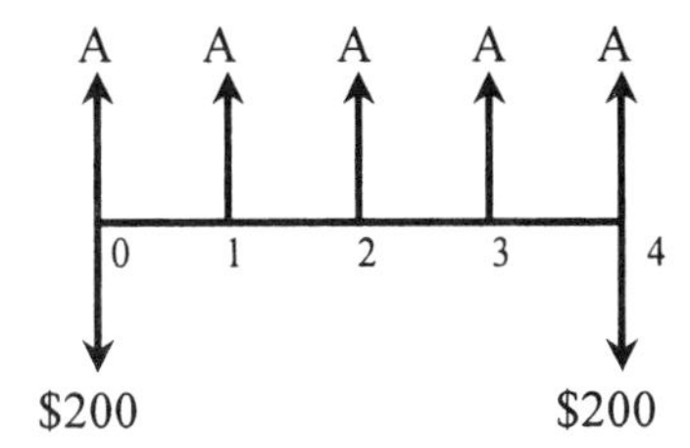

If i = 20%, find A.

Solution

$$F = \$200 + \$200(F/P, 20\%, 4) = \$614.80$$
$$A = \$614.80(A/F, 20\%, 5) = \$82.63$$

6-39

A(n)(fill in your major) student bought a car at a used car lot on 4th street for $2,000, including tax and insurance. He was to pay for the car by making 19 equal monthly payments, with the first payment to be made when the car was delivered (a down payment). Interest on the loan was charged at the rate of 12% compounded monthly. After 11 payments (the down payment and 10 end-of-month payments) were made, a second buyer agreed to buy the car from the student and to pay a cash amount to pay off the loan in full at the time the next payment was due. If there is no pay off penalty for the early pay off, what amount will be required to pay off the loan?

Solution

$$P = A(P/A, 1\%, 18) + A$$
$$2{,}000 = A(17.398269)$$
$$A = 2{,}000/17.398269$$
$$= \$114.95$$

$$\text{Payoff} = 114.95(P/A, 1\%, 7) + 114.95$$
$$= 114.95(6.728) + 114.95$$
$$= \$888.33$$

6-40

Data for tractors A and B are listed below. With interest of 12%, which tractor would be selected based on equivalent uniform annual cost (EUAC)?

	A	B
First cost	$30,000	$36,000
Annual maintenance	1,500	2,000
Salvage value	5,000	8,000

Useful life	6 years	6 years

Solution

S = Salvage value
A = Annual maintenance
P = First cost
i = 12 %
n = 6

EUAC = P(A/P, i %, n) - S(A/F, i %, n) + Other Costs

TRACTOR A:

EUAC = 30,000(A/P, 12%, 6) - 5,000(A/F, 12%, 6) + 1,500
= $8,180

check: EUAC = (P - S)(A/P, i %, n) + Si + Other Costs
= $8,180

TRACTOR B:

EUAC = 36,000(A/P, 12%, 6) - 8,000(A/F, 12%, 6) + 2,000
= $9,770

check: EUAC = (P - S)(A/P, 12%, 6) + S(.12) + Other Costs
= $9,770

Since criteria is to minimize EUAC select tractor A

6-41

According to the manufacturers' literature, the costs of running automatic grape peelers, if maintained according to the instruction manuals, are:

Manufacturer:	Slippery	Grater
First Cost	$500	$300
Maintenance	$100 at end of years 2, 4, 6 and 8	Year 1 - $ 0 2 - 50 3 - 75 4 - 100 5 - 125
Useful Life	10 years	5 years

Which alternative is preferred if MARR = 15%?

Solution

Slippery :

EUAC = [500 + 100(A/F, 15%, 2)(P/A, 15%, 8)](A/P, 15%, 10)
= $141.24

Grater:

EUAC = [300 + 25(P/G, 15%, 5) + 25(P/A, 15%, 4)(P/F, 15%, 1)](A/P, 15%, 5)
= $151.07

Therefore, choose slippery with lower EUAC.

6-42

A semiconductor manufacturer has been ordered by the city to stop discharging acidic waste liquids into the city sewer system. Your analysis shows you should select one of the following three systems.

System	Installed Cost	Annual Operating Cost	Salvage value End of 20 yrs
CleanH_2O	$30,000	$6,000	$2,000
AcidFree	35,000	5,000	5,000
Evergreen	80,000	1,000	40,000

If the system is expected to last and be used 20 years and money is worth 8%, which system should be purchased?

Solution

CleanH_2O EUAC = 6,000 + 30,000(A/P, 8%, 20) - 2,000(A/F, 8%, 20)
= $9,013

AcidFree EUAC = 5,000 + 35,000(A/P, 8%, 20) - 5,000(A/F, 8%, 20)
= $8,456

Evergreen EUAC = 1,000 + 80,000(A/P, 8%, 20) - 40,000(A/F, 8%, 20)
= $8,276

Purchase system with lowest EUAC, Evergreen

6-43

The following alternatives describe possible projects for the use of a vacant lot. In each case the project cost includes the purchase price of the land.

	Parking Lot	Gas Station
Investment Cost	$50,000	$100,000
Annual Income	35,000	85,000
Annual Operating Expenses	25,000	70,000 in Year 1, then increasing by 1,000/yr
Salvage Value	10,000	10,000
Useful Life	5 years	10 years

(a) If the minimum attractive rate of return (MARR) equals 18%, what should be done with the land?

(b) Is it possible the decision would be different if the MARR were higher than 18%? Why or why not? (No calculations necessary.)

Solution

(a) $EUAW_{P.L.}$ = (35,000 - 25,000) - 50,000(A/P, 18%, 5) + 10,000(A/F, 18%, 5) = -$4,592

$EUAW_{G.S.}$ = (85,000 - 70,000) - 100,000(A/P, 18%, 10) + 10,000(A/F, 18%, 10) - 1,000(A/G, 18%, 10) = -$10,019

Since both EUAW's are negative, leave lot vacant.

(b) No. Higher MARR favors lower cost projects and the lowest cost project(null) has already been chosen.

6-44

Given the following information about possible investments, what is the best choice at a minimum attractive rate of return (MARR) of 10%?

	A	B
Investment Cost	$5,000	$8,000
Annual Benefits	1,200	800
Useful Life	5 years	15 years

Solution

Equivalent annual worth is easier since the useful lives are different.

$EUAW_A$ = 1,200 - 5,000(A/P, 10%, 5) = -$119.00
$EUAW_B$ = 800 - 8,000(A/P, 10%, 15) = -$252.00

Although A is better than B, the Do-Nothing (Null) alternative is best.

6-45

You are considering purchasing the Press-o-matic or Steam-it-out model automatic ironing system to allow you to handle more dry cleaning business. Either machine will cost the same amount, $5,000.

The Press-o-matic will generate a positive cash flow of $1,300 per year for 5 years and then be of no service or salvage value.

The Steam-it-out will generate a positive cash flow of $800 per year for 10 years and then be of no service or salvage value.

You plan to be in the dry cleaning business for the next 10 years. How would you invest the $5,000 you have in your hand if you feel the time value of money is worth the same as your high interest bank account offers, which is

(a) 8%?
(b) 12%?

Solution

(a) Press EUAW = 1,300 - 5,000(A/P, 8%, 5) = $47.50
Steam EUAW = 800 - 5,000(A/P, 8%, 10) = $55.00

Choose highest EUAB, Steam-it-out

(b) Press EUAW = 1,300 - 5,000(A/P, 12%, 5) = -$87.00
Steam EUAW = 800 - 5,000(A/P, 12%, 10) = -$85.00

Choose neither option because both have a negative annual worth.

6-46

Data for Machines X and Y are listed below. With an interest rate of 8%, which machine would be selected based upon equivalent uniform annual cost (EUAC)?

	X	Y
First cost	\$5,000	\$10,000
Annual maintenance	500	200
Salvage value	600	1,000
Useful life	5 years	15 years

Solution

Method 1 EUAC = P(A/P, i %, n) - S(A/F, i %, n) + other costs
Method 2 EUAC = (P - S)(A/P, i %, n) + Si + other costs

Machine X:

Method 1 EUAC = 5,000(A/P, 8%, 5) - 600(A/F, 8%, 5) + 500 = \$1,650.20
Method 2 EUAC = (5,000 - 600)(A/P, 8%, 5) + 600 .08) + 500 = \$1,650.20

Machine Y:

Method 1 EUAC = 10,000(A/P, 8%, 15) - 1,000(A/F, 8%, 15) + 200 = \$1331.20
Method 2 EUAC = (10,000 - 1,000)(A/P, 8%, 5) + 1,000(.08) + 200 = \$1331.20

Decision criterion: minimize EUAC, therefore choose Y

6-47

Consider Projects A and B. Which project would you approve, if the income to both were the same. The expected period of service is 15 years, and the interest rate is 10%.

	Project A	Project B
Initial cost	\$50,000	\$75,000
Annual operating costs	15,000	10,000
Annual repair costs	5,000	3,000
Salvage value after 15 years	5,000	10,000

Solution

Project A:

$EUAC_A$ = 50,000(A/P, 10%, 15) + 20,000 – 5,000(A/F, 10%, 15) = \$26,417.50

Project B:

$EUAC_B = 75{,}000(A/P, 10\%, 15) + 13{,}000 - 10{,}000(A/F, 10\%, 15) = \$22{,}547.50$

Choose least cost: project B

6-48

Assuming a 10% interest rate, determine which alternative should be selected.

	A	B
First Cost	$5,300	$10,700
Uniform Annual Benefit	1,800	2,100
Useful Life	4 years	8 years
Salvage Value	0	200

Solution

Alternative A:

$$\begin{aligned} EUAW &= (5{,}300 - 0)(A/P, 10\%, 4) - 1{,}800 \\ &= \$127.85 \end{aligned}$$

Alternative B:

$$\begin{aligned} EUAW &= (10{,}700 - 200)(A/P, 10\%, 8) + 200(.1) - 2{,}100 \\ &= \$112.30 \end{aligned}$$

Choose alternative A

6-49

A company must decide whether to buy Machine A or Machine B. After 5 years A will be replaced with another A.

	Machine A	Machine B
First Cost	$10,000	$20,000
Annual Maintenance	1,000	0
Salvage Value	10,000	10,000
Useful Life	5 years	10 years

With the minimum attractive rate of return (MARR) = 10%, which machine should be purchased?

Solution

$EUAW_A$ = - 10,000(A/P, 10%, 5) - 1,000 + 10,000(A/F, 10%, 5) = -$2,000

$EUAW_B$ = - 20,000(A/P, 10%, 10) + 10,000(A/F, 10%, 10) = -$2,627

Therefore Machine A should be purchased.

6-50

The construction costs and annual maintenance costs of two alternatives for a canal are given below. Using equivalent uniform annual cost (EUAC) analysis, which alternative would you recommend? Assume 7% interest and infinite life. What is the capitalized cost of maintenance for the alternative you choose?

	Alternative A	Alternative B
Construction cost	$25,000,000	$50,000,000
Annual Maintenance costs	3,500,000	2,000,000

Solution

(a) Alternative A EUAC = A + P i = 3.5M + 25M(0.07) = $5,250,000
  Alternative B EUAC = A + P i = 2.0M + 50M(0.07) = $5,500,000

  Fixed Output ∴ minimize cost; choose A

(b) P = A/i = 3,500,000/0.07 = $50,000,000

6-51

Two alternatives are being considered by a food processor for the warehousing and distribution of its canned products in a sales region. These canned products come in standard cartons of 24 cans per carton. The two alternatives are:

Alternative A. To have its own distribution system.
The administrative costs are estimated at $43,000 per year, and other general operating expenses are calculated at $0.009 per carton. A warehouse will have to be purchased, which costs $300,000.

Alternative B. To sign an agreement with an independent distribution company, which is asking a payment of $0.10 per carton distributed.

Assume a study period of 10 years, and that the warehouse can be sold at the end of this period for $200,000.

(a) Which alternative should be chosen, if they expect that the number of cartons to be distributed will be 600,000 per year?

(b) Find the minimum number of cartons per year that will make the alternative of having a distribution system (Alt A.) more profitable than to sign an agreement with the distribution company (Alt B.)

Solution

(a) For 600,000 cartons/yr.

Alternative A :

Annual Costs:			
	Administration:		\$43,000
	Operating Expenses:	.009 x 600,000 =	5,400
	Capital Expenses*		36,270
		Total =	\$84,670

$$\text{*EUAC} = (P - S)(A/P, i, n) + S\,i$$
$$= (300{,}000 - 200{,}000)(A/P, 10\%, 10) + 200{,}000(0.1) = \$36{,}270$$

∴ Total annual costs = \$ 84, 670.00

Alternative B:

Total annual costs = 0.10 x 600,000 =\$ 60,000

∴ Sign an agreement for distribution (Alt. B)

(b) Let M = number of cartons/yr.

The EUAC for alternative B (agreement) = $EUAC_{AGREEMENT} = 0.10M$

The EUAC for alternative A (own system) = $EUAC_{OWN} = 43{,}000 + 0.009M + 36{,}270$

We want $EUAC_{OWN} < EUAC_{AGREEMENT}$

$$43{,}000 + .009M + 36{,}270 < 0.10M$$
$$79{,}270 < (0.10 - 0.009)M$$
$$79{,}270/0.091 < M$$
$$871{,}099 < M$$

∴ Own distribution is more profitable for 871,100 or more cartons/year.

6-52

The plant engineer of a major food processing corporation is evaluating alternatives to supply electricity to the plant. He will pay $3 million for electricity purchased from the local utility at the end of this first year and estimates that this cost will increase at $300,000 per year. He desires to know if he should build a 4000 kilowatt power plant. His operating costs (other than fuel) for such a power plant are estimated to be $130,000 per year. He is considering two alternative fuels:

(a) WOOD. Installed cost of the power plant is $1200/kilowatt. Fuel consumption is 30,000 tons per year. Fuel cost for the first year is $20 per ton and is estimated to increase at a rate of $2 per ton for each year after the first. No salvage value.

(b) OIL. Installed cost is $1000/kw. Fuel consumption is 46,000 barrels per year. Fuel cost is $34 per barrel for the first year and is estimated to increase at $1/barrel per year for each year after the first. No salvage value.

If interest is 12%, and the analysis period is 10 years, which alternative should the engineer choose? Solve the problem by equivalent uniform annual cost analysis (EUAC).

Solution

	Do Nothing	Wood	Oil
First Cost	0	4,000 x 1,200 = 4,800,000	4,000 x 1,000 = 4,000,000
Annual Oper. Costs	0	130,000	130,000
Annual Energy Costs	3,000,000	30,000 x 20 = 600,000	46,000 x 34 =1,564,000
Gradient	300,000	30,000 x 2 = 60,000	46,000 x 1 = 46,000

Do Nothing:

EUAC = 3,000K + 300K(A/G, 12%, 10) = $4,075,500

Wood:

EUAC = 4,800K(A/P, 12%, 10) + 130K + 600K + 60K(A/G, 12%, 10) = $1,794,700

Oil:

EUAC = 4,000K(A/P, 12%, 10) + 130K + 1,564K + 46K(A/G, 12%, 10) = $2,566,190

Minimize EUAC choose wood

6-53

The manager of F. Roe, Inc. is trying to decide between two alternative designs for an aquacultural facility. Both facilities produce the same number of fish for sale. The first alternative costs $250,000 to build, and has a first-year operating cost of $110,000. Operating costs are estimated to increase by $10,000 per year for each year after the first.

The second alternative costs $450,000 to build, and has a first-year operating cost of $40,000 per year, escalating at $5000 per year for each year after the first. The estimated life of both plants is 10 years and each has a salvage value that is 10% of construction cost.

Assume an 8% interest rate. Using equivalent uniform annual cost (EUAC) analysis, which alternative should be selected?

Solution

	Alternative 1	Alternative 2
First Cost	$250,000	$450,000
Uniform Annual Cost for 10 years	110,000	40,000
Gradient	10,000	5,000
Salvage in year 10	25,000	45,000

Alternative 1: EUAC = 250,000(A/P, 8%, 10) - 25,000(A/F, 8%, 10) + 110,000
+ 10,000(A/G, 8%, 10) = $184,235

Alternative 2: EUAC = 450,000(A/P, 8%, 10) - 45,000(A/F, 8%, 10) + 40,000
+ 5,000(A/G, 8%, 10) = $123,300

Fixed Output (same amount of fish for sale) ∴ Minimize EUAC

Choose Alternative 2

6-54

Two alternative investments are being considered. What is the minimum uniform annual benefit that will make Investment B preferable over Investment A? Assume interest is 10%.

Year	A	B
0	-$500	-$700
1-5	+150	?

Solution

$NPW_A = NPW_B$

$-500 + 150(P/A, 10\%, 5) = -700 + X(P/A, 10\%, 5)$
$X = \$202.76$

Alternate Solution:

$EUAW_A = EUAW_B$

$-500(A/P, 10\%, 5) + 150 = -700(A/P, 10\%, 5) + X$
$X = \$202.76$

6-55

Consider two investments:

(1) Invest $1000 and receive $110 at the end of each month for the next 10 months.

(2) Invest $1200 and receive $130 at the end of each month for the next 10 months.

If this were your money, and you want to earn at least 12% interest on your money, which investment would you make, if any? Solve the problem by annual cash flow analysis.

Solution

EUAW Analysis:

Alternative 1: EUAB - EUAC = 110 - 1,000(A/P, 1%, 10) = $4.40
Alternative 2: EUAB - EUAC = 130 - 1,200(A/P, 1%, 10) = $3.28

Maximum EUAB – EUAC, therefore choose alternative A.

Chapter 7

Rate of Return Analysis

7-1

Tony invested \$15,000 in a high yield account. At the end of 30 years he closed the account and received \$539,250. Compute the effective interest rate he received on the account.

Solution

Recall that $F = P(1 + i)^n$

$$539{,}250 = 15{,}000(1 + i)^{30} \Rightarrow 539{,}250/15{,}000 = (1 + i)^{30}$$
$$35.95 = (1 + i)^{30}$$
$$\sqrt[30]{35.95} = 1 + i$$
$$1.1268 = 1 + i$$
$$.1268 = i$$
$$i = 12.68\%$$

7-2

The heat loss through the exterior walls of a processing plant is estimated to cost the owner \$3,000 next year. A salesman from Superfiber, Inc. claims he can reduce the heat loss by 80 % with the installation of \$ 15,000 of Superfiber now. If the cost of heat loss rises by \$200 per year, after next year (gradient), and the owner plans to keep the building ten more years, what is his rate of return, neglecting depreciation and taxes?

Solution

NPW = 0 at the rate of return

Try 12%

$$NPW = -15{,}000 + .8(3{,}000)(P/A, 12\%, 10) + .8(200)(P/G, 12\%, 10)$$
$$= \$1{,}800.64$$

Try 15%

$$NPW = -\$237.76$$

By interpolation i = 14.7%

7-3

Does the following project have a positive or negative rate of return? Show how this is known to be true.

Investment cost	\$2,500
Net benefits	\$300 in Year 1, increasing by \$200/year
Salvage	\$50
Useful life	4 years

Solution

Year	Benefits
1	300
2	500
3	700
4	900
5	50
	Total = 2,450 < Cost

Total Benefits obtained are less than the investment, so the "return" on the investment is negative

7-4

At what interest rate would \$1000 at the end of 1995 be equivalent to \$2000 at the end of 2002?

Solution

$(1 + i)^7 = 2$; $i = (2)^{1/7} - 1 = 0.1041$ or 10.41%

7-5

A painting, purchased one month ago for \$1000, has just been sold for \$1700. What nominal annual rate of return did the owner receive on his investment?

Solution

i = 7,000/1,000 = 70% IRR = 70 x 12 = 840%

7-6

Find the rate of return for a \$10,000 investment that will pay \$1000/year for 20 years.

Solution

$10,000 = 1,000(P/A, i\%, 20)$

$(P/A, i\%, 20) = 10$

From tables: $7\% < i < 8\%$ $\therefore$ interpolate $i = 7.77\%$

7-7

A young engineer has a mortgage loan at a 15% interest rate, which he got some time ago, for a total of $52,000. He has to pay 120 more monthly payments of $620.72 each. As interest rates are going down, he inquires about the conditions under which he could refinance the loan. If the bank charges a new loan fee of 4% of the amount to be financed, and if the bank and the engineer agree on paying this fee by borrowing the additional 4% under the same terms as the new loan, what percentage rate would make the new loan attractive, if the conditions require him to repay it in the same 120 payments?

Solution

The amount to be refinanced:

a) PW of 120 monthly payments left

P = A(P/A, i, n)
= 620.72(P/A, 15%/12, 120)
= $38,474.09

b) New loan fee (4 %)

38,474.09 x 0.04 = $1,538.96

⇒ Total amount to refinance = 38,474.09 + 1,538.96 = 40,013.05

The new monthly payments are: A_{NEW} = 40,013.05 (A/P, i, 120)
The current payments are: A_{OLD} = 620.72

We want $A_{NEW} < A_{OLD}$

Substituting ⇒ 40,013.05(A/P, i, 120) < 620.72
(A/P, i, 120) < 620.72/40,013.05 = 0.0155

for i = 1% (A/P, 1%, 120) = 0.0143
for i = 1¼% (A/P, 1¼%, 120) = 0.0161

1% < i < 1¼% ∴ interpolate

i = 1.1667 %

This corresponds to a nominal annual percentage rate of 12 x 0.011667 = 14 %
Therefore, she has to wait until interest rates are less than 14%.

7-8

Your company has been presented with an opportunity to invest in a project. The facts on the project are presented below:

Investment required	$60,000,000
Annual Gross income	20,000,000
Annual Operating costs:	
Labor	2,500,000
Materials, licenses, insurance, etc.	1,000,000
Fuel and other costs	1,500,000
Maintenance costs	500,000
Salvage value after 10 years	0

The project is expected to operate as shown for ten years. If your management expects to make 25% on its investments before taxes, would you recommend this project?

Solution

$$\text{Net income} = 20M - (2.5M + 1M + 1.5M + .5M) = \$14{,}500{,}000$$

NPW = 0 at the rate of return

$$0 = -60{,}000{,}000 + 14{,}500{,}000(P/A, i, 10)$$
$$(P/A, i, 10) = 60/14.5 = 4.138$$

@20% P/A = 4.192
@25% P/A = 3.571

$20\% < i < 25\%$ ∴ interpolate

$i = 20.43\%$

IRR < 25 % ∴ Reject Project

7-9

Consider the following investment in a piece of land.

Purchase price	$10,000
Annual maintenance:	100
Expected sale price after 5 years:	$20,000

Determine:

(a) A trial value for i
(b) The rate of return (to 1/100 percent)
(c) What is the lowest sale price the investor should accept if she wishes to earn a return of 10% after keeping the land for 10 years?

Solution

(a) (F/P, i %, 5) = 20,000/10,000
= 2
i = 15 %

(b) NPW = -10,000 - 100(P/A, i %, 5) + 20,000(P/F, i %, 5) = 0

Try i = 15%: = -391.2
Try i = 12%: = +987.5

15% < i < 12% ∴ interpolate

i = 14.15 %

(c) NFW = 0 = -10,000(F/P, 10%, 10) - 100 (F/A, 10%, 10) + Sale Price

Sale Price = $27,534

7-10
Calculate the rate of return of the following cash flow with accuracy to the nearest 1/10 percent.

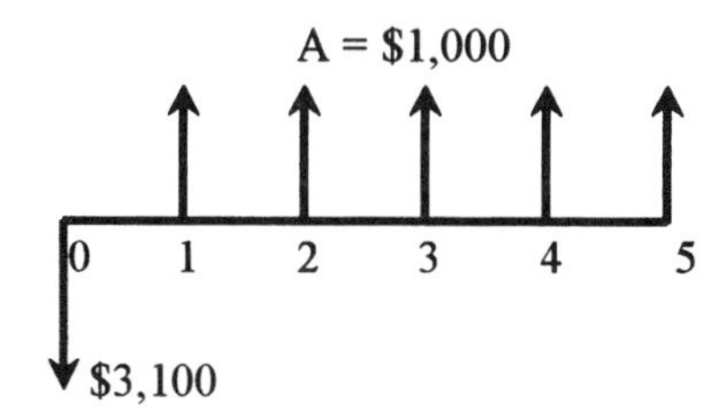

Solution

NPW = 0 at the rate of return

0 = -3,100 + 1,000(P/A, i, 5)
(P/A, i, 5) = 3.1

(P/A, 20%, 5) = 2.991
(P/A, 18%, 5) = 3.127

18% < i < 20% ∴ interpolate

i = 18.4%

7-11

An investment that cost $1000 is sold five years later for $1261. What is the nominal rate of return on the investment (interest rate)?

Solution

$$
\begin{aligned}
F &= P\,(F/P, i\,\%, 5)\\
1{,}261 &= 1{,}000\,(F/P, i\,\%, 5)\\
(F/P, i\,\%, 5) &= 1{,}261/1{,}000\\
&= 1.2610
\end{aligned}
$$

From tables:

$(F/P, 4\tfrac{1}{2}\%, 5) = 1.246$

$(F/P, 5\%, 5) = 1.276$

$4\tfrac{1}{2}\% < i < 5\% \quad \therefore$ interpolate

$i = 4.75\%$

7-12

Elizabeth made an initial investment of $5000 in a trading account with a stock brokerage house. After a period of 17 months the value of the account had increased to $8400. What is the nominal annual interest rate realized on the initial investment if it is assumed there were no additions or withdrawals from the account?

Solution

$$
\begin{aligned}
F &= P(F/P, i, 17)\\
F/P &= 8{,}400/5{,}000\\
&= 1.68
\end{aligned}
$$

$$
\begin{aligned}
(1 + i)^{17} &= 1.68\\
1 + i &= (1.68)^{1/17}\\
&= 1.031
\end{aligned}
$$

Annual interest rate = 3.1 x 12 = 37.2%

7-13

You borrowed $25 from a friend and after five months repaid $27. What interest rate did you pay (annual basis)?

Solution

$$F = P(1 + i)^n$$
$$27 = 25(1 + i)^5$$
$$(27/25)^{1/5} = 1 + i$$
$$= 1.0155$$

i = 1.55% per month or 18.6% per year

7-14

You have a choice of $3,000 now, or $250 now with $150 a month for two years. What interest rate will make these choices comparable?

Solution

3,000 = 250 + 150 (P/A, i, 24)
P/A = 18.33

@ 2% P/A = 18.914
@ 2½% P/A = 17.885

2% < i < 2½% ∴ interpolate

i = = 2.28% or 27.36% per year

7-15

Sain and Lewis Investment Management Inc., is considering the purchase of a number of bonds to be issued by Southeast Airlines. The bonds have a face value of $10,000 and a face rate of 7.5% payable annually. The bonds will mature ten years after they are issued. The issue price is expected to be $8,750. Determine the yield to maturity (IRR) for the bonds. If SLIM Inc. requires at least a 10% return on all investments should they invest in the bonds?

Solution

Year		10%	12%
0	First Cost	-8,750.00	-8,750.00
1-10	Interest 750(P/A, i%, 10)	+4,608.75	+4,813.50
10	Maturity 10,000(P/F, i% n)	+3,855.00	+4,224.00
	NPW =	-$286.25	$287.50

10% < IRR < 12% ∴ interpolate

i = 9.5% ⇒ Do not invest.

7-16

A 9.25% coupon bond issued by Gurley Gears LLC is purchased January 1, 2002 and matures December 31, 2010. The purchase price is $1079 and interest is paid semi-annually. If the face value of the bond is $1,000, determine the effective internal rate of return.

Solution

n = 9 x 2 = 18 ½year periods

½ Year		4%
0	First Cost	-1,079.00
1-18	Interest 46.25(P/A, i%, 18)	+585.48
18	Maturity 1000(P/F, i%, 18)	+493.60
	NPW =	$.08

$IRR = (1 + .04)^2 - 1$
$= 8.16\%$

7-17

Joe's Billiards Inc. stock can be purchased for $14.26 per share. If dividends are paid each quarter at a rate of $0.16 per share, determine the effective *i* if after 4 years the stock is sold for $21.36 per share.

¼ Year		3%	3½
0	Cost	-14.26	-14.26
1-16	Dividends .16(P/A, i%, 16)	+2.01	+1.93
16	Sale 21.36(P/F, i%, 16)	+13.31	+12.32
	NPW =	$1.06	-$.01

i = 3½ per quarter

effective $i = (1 + .035)^4 - 1$
$= 14.75\%$

7-18

A bond with a face value of $1,500 can be purchased for $800. The bond will mature five years from now and the bond dividend rate is 12%. If dividends are paid every three months, what effective interest rate would an investor receive if she purchases the bond?

Solution

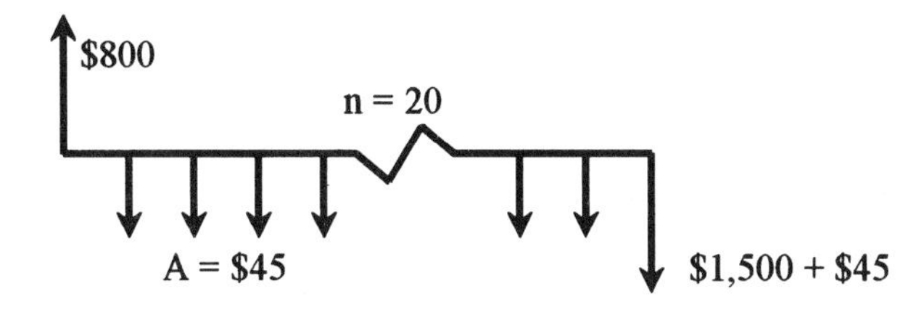

NPW = 0 at IRR

0 = -800 + 45(P/A, i%, 20) + 1,500(P/F, i%, 20)

By trial and error:

Try 7% NPW = +$64.33
Try 8% NPW = -$36.44

7% < i < 8% ∴ interpolate

i = 7.64 %

Effective interest rate = $(1+i)^m - 1 = (1+0.07638)^4 - 1 = 0.3423 = 34.23\%$

Chapter 7A

Difficulties Solving for an Interest Rate

7A-1

How many positive rates of return does the following cash flow have?

Year	Cash Flow
0	-$50,000
1	+25,000
2	+25,000
3	0
4	-50,000
5	+25,000
6	+25,000
7	+25,000

Solution

Year	Accumulated Cash Flow	
0	-$50,000	3 sign changes in cash flow
1	-25,000	
2	0	Cash Flow Rule of Signs:
3	0	There may be as many as 3 positive ROR
4	-50,000	
5	-25,000	Accumulated Cash Flow: $A_7 \neq 0$,
6	0	one sign change in accumulated cash flow,
7	$+25{,}000 = A_7$	$\therefore$ one positive rate of return

Note: also check to see if there is any external investments during project life.

7A-2

Consider the following cash flow generated by an investment opportunity presented to Insane Airlines.

Year	Cash Flow
0	$-800,000
1	-150,000
2	250,000
3	-100,000
4	400,000
5-10	250,000

If external investments yield 4%, determine the rate of return on the internal investment.

Solution

Year	Cash Flow		Transformed Cash Flow	NPW @ 12%	NPW @ 15%
0	$-800,000		-800,000	-800,000	-800,000
1	-150,000		-150,000	-133,935	-130,440
2	+250,000	-9,6154(1.04)↓	+153,846	122,646	116,323
3	-100,000	+100,000	0	0	0
4	+400,000		+400,000	254,200	228,720
5-10	+250,000		+250,000	653,135	540,923
				96,046	-44,474

By interpolation the rate of return is 14.05%

7A-3

If the external investment rate is 5%, determine the internal rate of return for the following cash flow:

Year	Cash Flow
0	-1,000
1	250
2	-250
3-5	250

Solution

Year	Cash Flow		Transformed Cash Flow	NPW @ 8%	NPW @ 9%
0	-1,000		-1,000	-1,000	-1,000
1	250	-238.10(1.05)↓	11.90	11.02	10.92
2	-250	+250	0	0	0
3-5	450		450	994.23	958.73
				5.25	-30.35

By interpolation, the rate of return is 8.15%.

Chapter 8

Incremental Analysis

8-1

Sugar-N-Spice, a cookie factory, needs a new cookie cutter machine. They have narrowed their choices to three different types of machines, which are presented below.

	A	B	C
Cost	\$18,000	\$25,000	\$15,000
Annual Net Savings	1,055	2,125	1,020
IRR	7%	9%	8%

Each choice has a twenty-five year useful life with no salvage value. If the MARR for Sugar-N-Spice is 8%, which choice should be chosen?

Solution

B-A

$\frac{25{,}000-18{,}000}{2125-1055}$ = P/A = 6.5421 = 12%~15% Choose most expensive, B

B-C

$\frac{25{,}000-15{,}000}{2125-1020}$ = P/A = 9.0498 = 8%~9% Choose most expensive, B

8-2

The construction firm of Cagle, Owens, and Wells (COW) is considering investing in newer, more productive forming equipment. The data concerning the three best alternatives is presented below.

Initial Investment	\$50,000	\$22,000	\$15,000
Annual Net Income	5,093	2,077	1,643
Computed IRR	8%	7%	9%

Each alternative has a twenty year useful life with no salvage value. If the MARR for COW is 7%, which alternative should be chosen? (Show ranges for IRR were appropriate.)

Solution

Incremental analysis is required for IRR.

A-B

NPW = 0 at IRR.

$$0 = [-50{,}000 - (-22{,}000)] + (5{,}093 - 2{,}077)(P/A, i\%, 20)$$

$$(P/A, i\%, 20) = 28{,}000/3{,}016 = 9.284$$

$8\% < IRR < 9\%$ Choose most expensive, A

A-C

$$(P/A, i\%, 20) = 35{,}000/3{,}450 = 10.145$$

$7\% < IRR < 8\%$ Choose most expensive, A

8-3

Barber Brewing is considering investing in one of the following opportunities.

	A	B	C	D	E
First Cost	$100.00	$130.00	$200.00	$330.00	$000.00
Annual Income	100.00	90.78	160.00	164.55	00.00
Annual Cost	73.62	52.00	112.52	73.00	00.00

Each alternative has a five year useful life. The firm's MARR is 8%. Which alternative should be selected?

Solution

NPW = 0 at IRR.

$0 = -\text{First cost} + \text{net income } (P/A, i\%, n)$ Therefore $(P/A, i\%, n) = \text{First cost/net income}$

IRR_A	$P/A = 100/26.38 = 3.791$	IRR = 10%	
IRR_B	$P/A = 130/38.78 = 3.352$	IRR = 15%	
IRR_C	$P/A = 200/47.48 = 4.212$	IRR = 6%	Eliminate, does not meet MARR
IRR_D	$P/A = 330/91.55 = 3.605$	IRR = 12%	
IRR_E		IRR = 0%	Eliminate, does not meet MARR

Incremental analysis required

IRR_{D-B} $P/A = 200/52.77 = 3.790$ IRR = 10% Choose D

IRR_{D-A} $P/A = 230/65.17 = 3.529$ $12\% < IRR < 15\%$ Choose D

8-4

The construction firm of Burns, Owens, Wells, and Lowrance LLC must replace a piece of heavy earth moving equipment. The information concerning the two best alternatives is summarized below. Each alternative is expected to last six years. If BOWL LLC has a cost of capital of 11%, which alternative should be chosen? Use IRR analysis.

	Big Bite	Dip Big
First Cost	$15,000	$22,500
Annual Operating Cost	3,000	1,500
Salvage Value	2,000	4,000

Solution

Since individual IRR's cannot be calculated perform incremental analysis.

DD - BB

Year		10%	12%
0	First Cost	-7,500.00	-7,500.00
1-6	Cost Savings 1500(P/A, i%, 6)	6,532.50	6,166.50
6	Salvage Value 2000(P/F, i%, 6)	1,129.00	1,013.20
		$161.50	-$320.30

10% < IRR < 12% Interpolation is required to determine if IRR > 11%.

IRR = 10.67% Therefore choose the least expensive, Big Bite.

8-5

Horizon Wireless must rebuild a cell tower recently destroyed on a tornado. If made of normal steel, the tower will cost $30,000 to construct and should last for 15 years. Maintenance will cost $1,000 per year. If a more corrosion resistant steel is used, the tower will cost $36,000 to build and the annual maintenance cost will be reduced to $250 per year. Determine the IRR of building the corrosion resistant bridge. If Horizon requires a return of 9% on its capital projects, which tower should they build?

Solution

$NPW_{NS} = -30{,}000 - 1000(P/A, i\%, 15)$
$NPW_{CRS} = -36{,}000 - 250(P/A, i\%, 15)$

$-30{,}000 - 1000(P/A, i\%, 15) = -36{,}000 - 250(P/A, i\%, 15)$
$6000 - 750(P/A, i\%, 15) = 0$
$(P/A, i\%, 15) = 8.000$

From tables 9% P/A = 8.061
10% P/A = 7.606

9% < IRR < 10% Interpolation yields IRR = 9.13%

Build the tower using corrosion resistant steel (the most expensive)

Chapter 9

Other Analysis Techniques

BENEFIT COST

9-1

Rash, Riley, Reed, and Rogers Consulting has a contract to design a major highway project that will provide service from Memphis to Tunica, Mississippi. R^4 has been requested to provide an estimated B/C ratio for the project. Relevant data are:

Initial Cost	\$20,750,000
Right of way maintenance	550,000
Resurfacing (every 8 years)	10% of first cost
Shoulder grading and re-work (every 6 years)	750,000
Average number of road users per year	2,950,000
Average time savings value per road user	\$2

Determine the B/C ratio if i = 8%.

Solution

$$AW_{BENEFITS} = 2{,}950{,}000 \times \$2 = \$5{,}840{,}000$$
$$AW_{COSTS} = 20{,}750{,}000(A/P, 8\%, \infty) + 550{,}000 + .10(20{,}750{,}000)(A/F, 8\%, 8) + 750{,}000(A/F, 8\%, 6) = \$2{,}507{,}275$$

$$B/C = \frac{AW_{BENEFITS}}{AW_{COSTS}} = \frac{5{,}840{,}000}{2{,}507{,}275} = 2.33$$

9-2

A proposed bridge on the interstate highway system is being considered at the cost of \$2 million. It is expected that the bridge will last 20 years. Operation and maintenance costs are estimated to be \$180,000 per year. The federal and state governments will pay these construction costs. Benefits to the public are estimated to be \$900,000 per year. The building of the bridge will result in an estimated cost of \$250,000 per year the general public. The project requires a 10% return. Determine the B/C ratio for the project. State any assumption made about benefits or costs.

Solution

$250,000 cost to general public is disbenefit.

$AW_{BENEFITS}$ = 900,000 - 250,000 = $650,000
AW_{COSTS} = 2,000,000(A/P, 10%, 20) + 180,000 = $415,000

$$B/C = \frac{AW_{BENEFITS}}{AW_{COSTS}} = \frac{650,000}{415,000} = 1.57$$

9-3

The town of Podunk is considering the building of a new downtown parking lot. The land will cost $25,000 and the construction cost of the lot is estimated to be $150,000. Each year costs associated with the lot are estimated to be $17,500. The income from the lot is estimated to be $18,000 the first year and increase by $3,500 each year for the twelve year expected life of the garage. Determine the B/C ratio if Podunk uses a cost of capital of 4%.

Solution

$PW_{BENEFITS}$ = 18,000(P/A, 4%, 12) + 3500(P/G, 4%, 12) = $334,298
PW_{COSTS} = 175,000 + 17,500(P/A, 4%, 12) = 339,238

$$B/C = \frac{PW_{BENEFITS}}{PW_{COSTS}} = \frac{334,298}{339,238} = 0.99$$

9-4

Flyin' Ryan's Tire Sales is considering the purchase of new tire balancing equipment. The machine will cost $12,699 and have an annual savings of $1,500 with a salvage value at the end of 12 yrs. of $250. If MARR is 6%, use B/C analysis to determine whether or not Ryan should purchase the machine.

Solution

$PW_{BENEFITS}$ = $1500(P/A, 6%, 12) + $250(P/F, 6%, 12) = $12,700.25
PW_{COSTS} = $12,699

$$B/C = \frac{PW_{BENEFITS}}{PW_{COSTS}} = \frac{12,700}{12,699} = 1.00$$

Conclusion: Yes, the machine should be purchased

9-5

Sim City wants to build a new bypass between two major roads that will cut travel time for commuters. The road will cost $2,000,000 and save 7,500 people $100/yr in gas. The road will need to be resurfaced every year at a cost of $7,500. The road is expected to be in use for 20 years. Determine if Sim City should build the road using B/C analysis. MARR is 8%.

Solution

PW of Costs 2,000,000 + 7,500(P/A, 8, 20) = $2,073,635
PW of Benefits (7,500)($100)(P/A, 8, 20) = $7,363,500

B/C = 7,363,500/2,073,635 = 3.551

Conclusion: Yes, Sim City should build the bypass

FUTURE WORTH

9-6

A mortgage of $50,000 for 30 years, with monthly payments at 10% interest is contemplated. At the last moment you receive news of a $25,000 gift from you parents to be applied to the principal. Leaving the monthly payments the same, what amount of time will now be required to pay off the mortgage and what is the amount of the last payment (assume any residual partial payment amount is added to the last payment)?

Solution

i = 6%/12 = ½% per month   n = 30 x 12 = 360 periods

A = 50,000(A/P, ½%, 360)
= $299.77 monthly payment   (Note: For more accurate answer $\frac{1}{P/A}$ factor was used)

After reduction of P to 25,000

25,000 = 299.77(P/A, i%, n)
(P/A, i%, n) = 83.40

Try n = 104 periods; P/A = 80.942
Try n = 120 periods: P/A = 90.074

Interpolate, n = 108.31 periods = 9.03 years

At 9 years (108 periods): P = 299.77 (P/A, ½%, 108)
= 299.77 (83.2934)
= 24,968.87

Residual = 25,000 - 24,968.87 = $31.13

Last Payment = Value of residual at time of last payment + last payment
= 31.13(F/P, ½%, 108) + 299.77
= $353.12

9-7

A woman deposited $10,000 into an account at her credit union. The money was left on deposit for 10 years. During the first five years the woman earned 9% interest (nominal), compounded monthly. The credit union then changed its interest policy so that the second five years the woman earned 6% interest (nominal), compounded quarterly.

(a) How much money was in the account at the end of the 10 years?
(b) Calculate the rate of return that the woman received.

Solution

(a) at the end of 5 years:
F = 10,000 (F/P, ¾%, 60)* = $15,660.00
at the end of 10 years:
F = 15,660(F/P, ½%. 20)** = $17,304.30

(b) 10,000(F/P, i, 10) = 17,304.30
(F/P, i, 10) = 1.7304

try i = 5% (F/P, i, 10) = 1.629
try i = 6% (F/P, i, 10) = 1.791

5% < i < 6% ∴ interpolate

i = 5.64%

* i = 9%/12 n = 5 x 12
** i = 6%/4 n = 5 x 4

9-8

A woman deposited $100 per month in her savings account for 24 months at 6% interest, compounded monthly. Then for five years she made no deposits or withdrawals. How much is the account worth after seven years?

Solution

$i = 6\%/12 = ½\%$ per month

$FW = 100\ (F/A, ½\%, 24)\ (F/P, ½\%, 60)$
$= \$3,430.78$

9-9

On January 1st a sum of $1,200 is deposited into a bank account that pays 12% interest, compounded monthly. On the first day of each succeeding month $100 less is deposited (so $1,100 is deposited February 1st, $1,000 on March 1st, and so on). What is the account balance immediately after the December 1st deposit is made?

Solution

$P = 1,200(P/A, 1\%, 12) - 100\ (P/G, 1\%, 12)$
$= 7,449.1$
$F = 7,449.10(F/P, 1\%, 12)$
$= \$8,395.14$

9-10

A 25-year-old engineer named Newt begins working for a salary of $50,000 per year when he graduates from college. From his first monthly paycheck, he notices 7% of his salary is deducted and paid into Social Security, and his employer pays a like amount. In effect, Newt finds that 14% of his salary is being taken by the government for this mandatory program.
Assuming that Newt's contribution to Social Security is 14% of $50,000 = $7,000 per year, and that Newt works for the same salary until he is 65, how much will he have effectively contributed into the Social Security program. Assume a 5% interest rate.

Solution

$F = 7,000(F/A, 5\%, 40)$
$= \$845,593.00$

9-11

How much money would be in an account if $1,000 is deposited in a bank at 4% interest, compounded semiannually, for 3 years?

Solution

$i = 4\%/2 = 2\%$ per ½ year; $n = 2 \times 3 = 6$ periods

$F = \$1,000\ (F/A, 2\%, 6) = \$1126$

9-12

Harry was a big winner in the New Hampshire sweepstakes. After paying the income taxes he had \$80,000 left to invest in an investment fund that will pay 10% interest for the next 20 years. How much money will Harry receive at the end of the 20 years?

Solution

$$F = 80{,}000(F/P, 10\%, 20) = \$538{,}160.00$$

9-13

David has received \$20,000 and wants to invest it for 12 years. There are three plans available to him.

(a) A Savings Account. It pays 3¾% per year, compounded daily.

(b) A Money Market Certificate. It pays 6¾% per year, compounded semiannually.

(c) An Investment Account. Based on past experience it is likely to pay 8½% per year.

If David does not withdraw the interest, how much will be in each of the three investment plans at the end of 12 years?

Solution

a) $F = P(1 + i)^n$

$$i_{eff} = \left(1+\frac{r}{m}\right)^m - 1 = \left(1+\frac{.0375}{365}\right)^{365} - 1 = 3.82\%$$

$$FW = \$20{,}000(1 + .0382)^{12} = \$31{,}361.89$$

b) $$i_{eff} \left(1+\frac{.0675}{2}\right)^2 - 1 = 6.86\%$$

$$FW = \$20{,}000(1 + .0686)^{12} = \$44{,}341.67$$

c) $$FW = \$20{,}000(1 + 0.115)^{12} = \$73{,}846.24$$

Choose plan C since this plan yields the highest return at the end of 12 years.

9-14

How long will it take for \$300 to triple at a 5% per year interest rate?

Solution

Let F = 3 and P = 1

3 = 1(F/P, 5%, n)

3 = (F/P, 5%, n)

n	(F/P, 5%, n)	
22	2.925	
.22.5	3.000	By interpolation ≈ 22.5 years required to triple
23	3.072	

9-15
An annuity is established by the payment of $150 per month for eight years with interest to be calculated at 7½%. The company retains these funds in an account from which they propose to pay you $1,530 per year for life (an actuarial period of 18 years for your age). If interest is assumed to continue at 7½%, what is the lump-sum profit to the company at the end of the pay-in period? Assume monthly compounding for the payment period.

Solution

$$FW = 150(F/A, \frac{.075}{12}, 96) = 150(130.995) = \$19,649$$

$$i_{eff} \left(1+\frac{.075}{12}\right)^{12} - 1 = 7.76\%$$

PW of payment = 1,530(P/A, 7.76%, 18) = 1,530(9.530) = $14,581
Profit at end of pay-in period = 19,649 - 14,581 = $5,068

9-16
A person would like to retire 10 years from now. He currently has $32,000 in savings, and he plans to deposit $300 per month, starting next month, in a special retirement plan. The $32,000 is earning 8% interest, while the monthly deposits will pay him 6% nominal annual interest. Once he retires, he will collect the two sums of money, and being conservative in his calculations, he expects to get a 4% annual interest rate after year 10. Assuming he will only spend the interest he earns, how much will he collect in annual interest, starting in year 11?

Solution

Savings:

$$F = 32{,}000(F/P, 8\%, 10) = \$69{,}086$$

Monthly deposits:

$$F = 300(F/A, \frac{.06}{12}, 120) = \$49{,}164$$

The total amount on deposit at the end of year 10 is:

$$F_T = 69{,}086 + 49{,}164 = \$118{,}250$$

The interest to collect per year = 118,250 x 0.04 = 4,730

9-17

Mary wants to accumulate a sum of $20,000 over a period of 10 years to use as a downpayment for a house. She has found a bank that pays 6% interest compounded monthly. How much must she deposit four times a year (once each three months) to accumulate the $20,000 in ten years?

Solution

In this problem Mary's deposits do not match the interest period. One solution is to compute what her monthly deposit would need to be, and then the equivalent deposit each 3 months.

Monthly deposit (A) = F(A/F, i %, n) - 20,000(A/F, ½%, 120) = 20,000 (0.00610) = $122
Equivalent deposit at end of each 3-months: = A(F/A, ½%, 3) = 122(3.015) = $368
Note that the compound interest table only provided two digit accuracy.

<u>Hand Calculator solution</u>:

$$\text{monthly deposit (A)} = 20{,}000\left[\frac{0.005}{(1+0.005)^{120}-1}\right] = 122.041$$

$$\text{equivalent deposit at end of each 3-months} = 122.041\left[\frac{(1+0.005)^{3}-1}{0.005}\right] = \$367.96$$

9-18

If $5,000 is deposited into a savings account that pays 8% interest, compounded quarterly, what will the balance be after 6 years? What is the effective interest rate?

Solution

$$F = 5{,}000(F/P, 2\%, 24)$$
$$= \$8{,}042.19$$

9-19

Starting now, deposits of $100 are made each year into a savings account paying 6% compounded quarterly. What will be the balance immediately after the deposit made 30 years from now?

Solution

F = 100 (A/F, 1½%,4) (F/A, 1½%, 120) + 100 (F/P, 1½%, 120)
= $8,693.58

9-20

You invest $1,000 in a bank at 4% nominal interest. Interest is continuously compounded. What will your investment be worth

(a) at the end of 1 year?
(b) at the end of 36 months?

Solution

(a) $F = Pe^{rn} = 1{,}000e^{(0.04)(1)} = \$1{,}040.81$

(b) $F = Pe^{rn} = 1{,}000e^{(0.04)(3)} = \$1{,}127.50$

9-21

Using the tables for Uniform Gradients, solve for the Future Value at the end of year 7.

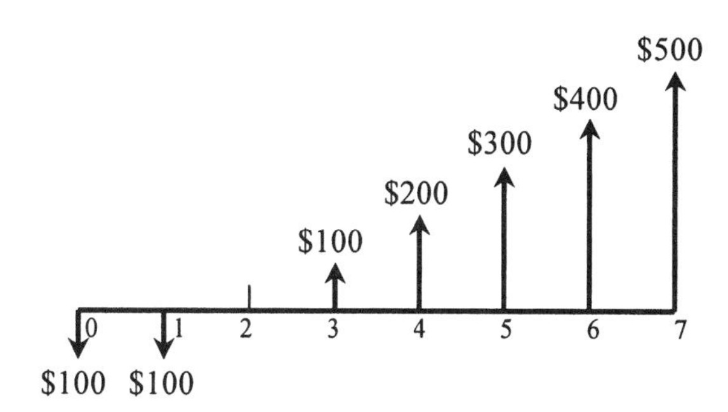

Solution

PV = 100(P/G, 10%, 7) - 100(P/A, 10%, 7) - 100
= $689.50

FV = 689.5(F/P, 10%, 7)
= $1,343.84

9-22

If Jane invests $125,000 in a fund that will pay 8%, compounded quarterly, how much will she have in the fund at the end of 10 years? What effective interest rate is she earning on her investment?

Solution

i = 8%/4 = 2% per quarter n = 10 x 4 = 40 periods

F = 125,000(F/P, 2%, 40) = $276,005

$$i_{eff} = \left(1+\frac{r}{m}\right)^m - 1 = \left(1+\frac{.08}{4}\right)^4 - 1 = 8.24\%$$

9-23

As a tax shelter, Joan's father has set up a trust fund into which he puts $1000 every year on Joan's birthday beginning on her first birthday. How much will be in the fund on Joan's 21st birthday if the account pays 6% compounded quarterly?

Solution

i = 6%/4 = 1½ per quarter

A = 1,000(A/F, 1½%, 4) = $244.40
F = 244.40(F/A, 1½%, 84) = $40,613.

9-24

An engineer is considering the purchase of a new set of batteries for a tractor. Given the cost, annual benefit and useful life, conduct a net future worth (NFW) analysis to decide which alternative to purchase if i = 12%.

	A	B
Cost	$150	$90
Annual benefit	40	40
Useful life	6 yrs	3 yrs

Solution

Alternative A:

NFW = 40(F/A, 12%, 6) - 150(F/P, 12%, 6) = $28.5

Alternative B:

NFW = 40(F/A, 12%, 6) - 90(F/P, 12%, 6) - 90(F/P, 12%, 3) = $20.49

Choose Alternative A, largest NFW

PAYBACK PERIOD

9-25

Two mutually exclusive alternatives are found to be acceptable but A lasts twice as long as B. With no additional information given, which alternative is likely to have the shorter payback period? Why?

Solution

Alternative B is likely to have the shorter payback period. Since both are acceptable, they must return benefits greater than costs within the useful life. But B has only half as long a life, so it will almost certainly return benefits greater than cost before A and will therefore have a shorter payback period.

9-26

What is the major advantage of using payback period to compare alternatives?

Solution

Payback period is easy and rapid. Also, it provides information on investment recovery time that may be important for companies with cash flow problems.

9-27

For calculating payback period, when is the following formula valid?

$$\text{Payback period} = \frac{\text{First Cost}}{\text{Annual Benefits}}$$

Solution

Valid when:
a) There is a single first cost at time zero.
b) Annual Benefits = Net annual benefits after subtracting any annual costs
c) Net Annual Benefits are uniform

9-28
Is the following statement True or False?

If two investors are considering the same project, the payback period will be longer for the investor with the higher minimum attractive rate of return (MARR).

Solution

Since payback period is generally the time to recover the investment, and ignores the MARR, it will be the same for both investors. The statement is False.

9-29
What is the payback period for a project with the following characteristics, if the minimum attractive rate of return (MARR) is 10%?

First Cost	$20,000
Annual Benefits	8,000
Annual Maintenance	2,000 in year 1, then increasing by $500/year
Salvage Value	2,000
Useful Life	10 years

Solution

Payback occurs when the sum of net annual benefits is equal to the first cost. Time value of money is ignored.

Year	Benefits	-	Costs	=	Net Benefits	Total Net Benefits
1	8,000	-	2,000	=	6,000	6,000
2	8,000	-	2,500	=	5,500	11,500
3	8,000	-	3.000	=	5,000	16,500
4	8,000	-	3.500	=	4.500	21,000>20,000

Payback period = 4 years (Actually a little less)

9-30
Determine the payback period (to the nearest year) for the following project if the MARR is 10%.

First Cost	$10,000
Annual Maintenance	500 in year 1, increasing by $200/year
Annual Income	3,000
Salvage Value	4,000
Useful Life	10 years

Solution

Year	Net Income	Sum
1	2,500	2,500
2	2,300	4,800
3	2,100	6,900
4	1.900	8,800
5	1,700	10,500 > 10,000

Payback period = 5 years

9-31
Determine the payback period (to the nearest year) for the following project:

Investment cost	$22,000
Annual maintenance costs	1,000
Annual benefits	6,000
Overhaul costs	7,000 every 4 years
Salvage Value	2,500
Useful life	12 years
MARR	10%

Solution

Year	Σ Costs	Σ Benefits	
0	22,000	--	
1	23,000	6,000	
2	24,000	12,000	
3	25,000	18,000	
4	33,000	24,000	
5	34,000	30,000	
6	35,000	36,000	← Payback

Payback period = 6 years

9-32

A cannery is considering different modifications to some of their can fillers in two plants that have substantially different types of equipment. These modifications will allow better control and efficiency of the lines. The required investments amount to $135,000 in Plant A and $212,000 for Plant B. The expected benefits(which depend on the number and types of cans to be filled each year) are as follows:

Year	Plant A Benefits	Plant B Benefits
1	$ 73,000	$ 52,000
2	73,000	85,000
3	80,000	135,000
4	80,000	135,000
5	80,000	135,000

(a) Assuming MARR = 10%, which alternative is should be chosen?
(b) Which alternative should be chosen based on payback period?

Solution

a) May be solved in various ways. Use PW method.

$$NPW_A = -135K + 73K(P/A, 10\%, 2) + 80K(P/A, 10\%, 3)(P/F, 10\%, 2)$$
$$= \$156,148.50$$

$$NPW_B = -212K + 52K(P/F, 10\%, 1) + 85K(P/F, 10\%, 2)$$
$$+ 135K(P/A, 10\%, 3)(P/F, 10\%, 2)$$
$$= \$182,976.80$$

Therefore, modifications to plant B are more profitable

b)

	PLANT A		PLANT B	
YEAR	BENEFITS	CUMMULATIVE BENEFITS	BENEFITS	CUMMULATIVE BENEFITS
1	73,000	73,000	52,000	52,000
2	73,000	146,000*	85,000	137,000
3	80,000	226,000	135,000	272,000**

*The PBP of A is less than 2 years (1.85 years)

**The PBP of B is less than 3 years (2.55 years)

Therefore, although not the most profitable, alternative A has the shortest payback period and should be chosen.

9-33

In this problem the minimum attractive rate of return is 10%. Three proposals are being considered.

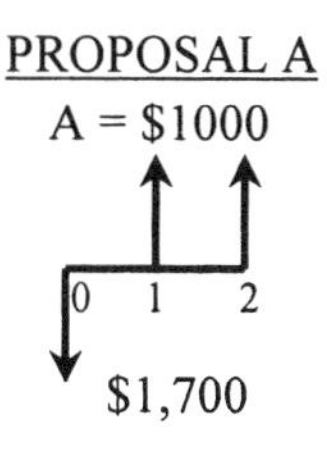

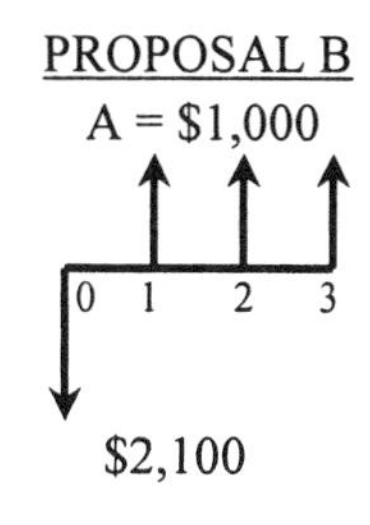

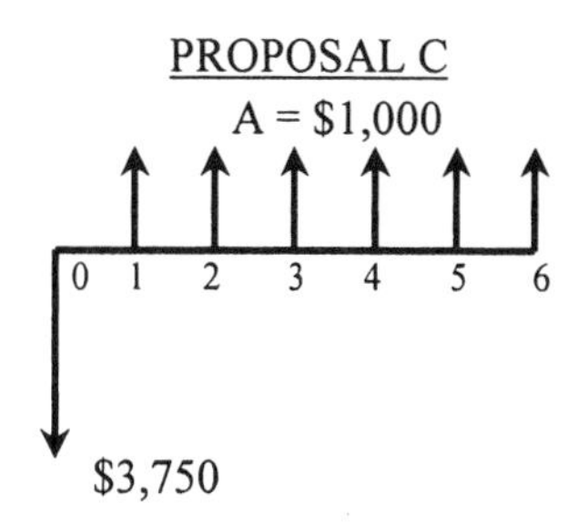

(a) Which proposal would you choose using Future Value analysis?
(b) How many years for Payback for each alternative?

Solution

(a) Proposal A EAB = 1,000 - 1,700(A/P, 10%, 2)
= \$20.50
Proposal B EAB = 1,000 - 2,100(A/P, 10%, 3)
= \$155.60
Proposal C EAB = 1,000 - 3,750(A/P, 10%, 6)
= \$139.00

Proposal A: FW = 20.50(F/A, 10%, 6) = \$ 158.20
Proposal B: FW = 155.60(F/A, 10%, 6) = \$1,200.60
Proposal C: FW = 139.00(F/A, 10%, 6) = \$1,072.50

Choose highest FW, Proposal B

(b) Proposal A: $\frac{1,700}{1,000} = 1.7$ years

Proposal B $\frac{2,100}{1,000} = 2.1$ years

Proposal C $\frac{3,750}{1,000} = 3.75$ years

SENSITIVITY ANALYSIS BREAKEVEN

9-34

A machine that produces a certain piece must be turned off by the operator after each piece is completed. The machine "coasts" for 15 seconds after it is turned off, thus preventing the operator from removing the piece quickly before producing the next piece. An engineer has suggested installing a brake that would reduce the coasting time to 3 seconds.

The machine produces 50,000 pieces a year. The time to produce one piece is 1 minute 45 seconds, excluding coasting time. The operator earns $8.00 an hour and other direct costs for operating the machine are $4.00 an hour. The brake will require servicing every 500 hours of operation. It will take the operator 30 minutes to perform the necessary maintenance and will require $44.00 in parts and material. The brake is expected to last 7500 hours of operation (with proper maintenance) and will have no salvage value.

How much could be spent for the brake it the minimum attractive rate of return is 10% compounded annually?

Solution

$$\text{Annual cost w/o the brake} = 50{,}000\left(\frac{2}{60}\right)(12) = \$20{,}000$$

$$\text{Annual cost w/ the brake} = 50{,}000\left(\frac{1.8}{60}\right)(12) = \$18{,}000$$

$$\text{Maintenance} = \left[\frac{\left(50{,}000\left(\frac{1.8}{60}\right)\right)}{500}\right]\left(.5(12+44)\right) = \$150$$

$$\text{Brake will last: } \frac{7{,}500}{\left(\frac{5{,}000(1.8)}{60}\right)} = 5 \text{ years}$$

Max. amount = (20,000 - 18,150)(P/A, 10%, 5) = $7,013.35

9-35

A road can be paved with either asphalt or concrete. Concrete costs $15,000/km and lasts 20 years. What is the maximum that should be spent on asphalt if it only lasts 10 years? Annual maintenance costs for both pavements are $500/km. MARR = 12%.

Solution

Since maintenance is the same for both, it doesn't affect the answer. However, there is nothing wrong with including it.

$$15{,}000(A/P, 12\%, 20) = P_{ASPHALT}(A/P, 12\%, 10)$$
$$15{,}000(.1339) = P_{ASPHALT}(.1770)$$
$$P_{ASPHALT} = \$11{,}347$$

9-36

A proposed building may be roofed in either galvanized steel sheet or composition roofing. The composition roof costs $20,000 and must be replaced every 5 years at the same cost. The steel roof costs $28,000 but the useful life is unknown. Neither roof has any salvage value and no maintenance is needed. If the minimum attractive rate of return (MARR) equals 15%, what is the minimum life that the steel roof must have to make it the better alternative? (Report to the nearest whole year; don't bother interpolating.)

Solution

$$EAC_S = EAC_C$$

$$28{,}000(A/P, 15\%, n) = 20{,}000(A/P, 15\%, 5)$$
$$28{,}000(A/P, 15\%, n) = 5{,}966$$

$$(A/P, 15\%, n) = .2131$$

$$(A/P, 15\%, 8) = .2229$$
$$(A/P, 15\%, 9) = .2096 \quad \rightarrow \quad n = 9$$

9-37

What is the breakeven capital cost for Project B compared to Project A if interest equals 10%?

Year	A	B
0	-1,000	?
1 - 5	+300/year	+200/year

Solution

$$\text{NPW of A} = -1{,}000 + 300\,(P/A, 10\%, 5)$$
$$= 137.3$$
$$\text{NPW of B} = \text{NPW of A}$$
$$137.3 = P_B + 200(P/A, 10, 5)$$
$$P_B = -\$620.90$$

9-38

What is the smallest acceptable annual income from a project that has a $70,000 investment cost and a $70,000 salvage value if the life is 15 years and the minimum attractive rate of return (MARR) is 20%?

Solution

Income = 70,000(A/P, 20%, 15) - 70,000(A/F, 20%, 15)
= 70,000[(A/P, 20%, 15) – (A/F, 20%, 15)]
= $14,000

9-39

A car rental agency has a contract with a garage to have them do major repairs (specified in the contract) for $450/car every six months. The car rental agency estimates that for $150,000, amortized at 8% interest for 20 years, and a salvage value of $60,000, they could have their own facility. They estimate that they could take care of their own car repairs in this facility at a cost of $200/car every six months. Ignoring taxes and other economic factors, what is the minimum number of cars needed to make the change feasible?

Solution

Let N = number of autos needed

450N = (150,000 - 60,000)(A/P, 4%, 40) + 60,000(.04) + 200N
450N = 90,000 x .0505 + 2,400 + 200N
250N = 6,945
N = 27.78 or 28 autos needed

9-40

Assume you need to buy some new automobile tires and you are considering purchasing either the "Econo-Ride", which costs $33.95 per tire, or the "Road King", which costs $65.50. Both tires are alike except that the "Road King" is more durable and will last longer. Regardless of which tire is purchased, balancing and installation costs are $1.50 per tire. The salesman says the "Econo-Ride" will last 20,000 miles. Assume a minimum attractive rate of return (MARR) of 6% and that you drive 10,000 miles per year.

(a) How many miles would the "Road King" have to last to make you indifferent in your choice?
(b) The salesman says the "Road King" will be on sale next week. If he also says the tire will last 30,000 miles, what would the sale price have to be to make you indifferent in your choice?

Solution

(a) $4(1.5 + 33.95)(A/P, 6\%, \frac{20{,}000}{10{,}000}) = 4(1.5 + 65.50)(A/P, 6\%, N)$

$(A/P, 6\%, N) = .28859$

From tables N = 4, or 40,000 miles

(b) $4(1.5 + 33.95)(A/P, 6\%, \frac{20{,}000}{10{,}000}) = 4(1.5 + P)(A/P, 6\%, \frac{30{,}000}{10{,}000})$

$P = \$50.18$

9-41

A soft drink company has researched the possibility of marketing a new low-calorie beverage, in a study region. The expected profits depend largely on the sales volume, and there is some uncertainty as to the precision of the sales-forecast figures. The estimated investment is $173,000 while the anticipated profits are $49,500 per year for the next 6 years. If the company's MARR = 15%, is the decision to invest sensitive to the uncertainty of the sales forecast, if it is estimated that in the worst case the profits will be reduced to $40,000 per year? What is the minimum volume of sales for the project to breakeven, if there is a profit of $6.70 per unit volume?

Solution

a) For an annual profit of $49,500
NPW = 49,500 (P/A, 15%, 6) - 173,000
= +14,308 (attractive)

b) For an annual profit of $40,000
NPW = 40,000 (P/A, 15%, 6) - 173,000
= -21,640 (not attractive)

Therefore the decision is sensitive to the expected variations in sales or profits.

The breakeven: NPW = 0
NPW = 0 = X(P/A, 15%, 6) - 173,000 where X = min $ profit

$$X = \frac{173{,}000}{(P/A, 15\%, n)} = \$45{,}718.80$$

$$\text{in volume units} = \frac{\$45{,}718.80}{\$6.70/\text{unit}} = 6{,}824 \text{ volume units}$$

9-42

A machine, costing $2,000 to buy and $300 per year to operate, will save labor expenses of $650 per year for 8 years. If the interest rate is 10%, what is the minimum salvage value (after 8 years) at which the machine is worth purchasing?

Solution

NPW 0 = - 2,000 + 350(P/A, 10%, 8) + S(P/F, 10%, 8)
0 = - 132.75 + .4665S
S = $284.57

9-43

The PARC Company can purchase gizmoes to be used in building whatsits for $90 each. PARC can manufacture their own gizmoes for $7,000 per year overhead cost plus $25 direct cost for each gizmoe, provided they purchase a gizmoe maker for $100,000. PARC expects to make whatsits using gizmoes for 10 years. The gizmoe maker should have a salvage value of $20,000 after 10 years. PARC uses 12% as its minimum attractive risk rate. At what annual production rate should PARC make their own gizmoes?

Solution

Equivalent Cost Solution:
EAC_{BUY} = $90N where N = annual quantity
EAC_{MAKE} = 100,000(A/P, 12%, 10) + 7,000 + 25N -20,000(A/F, 12%,10)
= 23,560 + 25N

For breakeven:
$EAC_{BUY} = EAC_{MAKE}$
90N = 23,560 + 25N
N = 362.5

This indicates they should be bought at 362/year or less and made at 363/year or more.

9-44

A coal burning power plant has been ordered by the government to install a $5 million pollution abatement device to remove sulphur that is currently being emitted into the air. The sulphur is removed by allowing the plant's exhaust to pass through a filter. The filtration system requires the presence of a certain chemical. The purchase price of the chemical is $1000 per kilogram. Studies have been conducted that show that the number of units of sulphur that may be recovered annually from the exhaust is equal to 100 times the square root of the number of kilograms of the chemical used in the filtration system.

Therefore:

$$(\text{units of sulphur}) = 100 \text{ x } (\text{kg of chemical})^{1/2}$$

Each unit of sulphur that is removed may then be sold by the power plant to chemical supply companies for $300. The filtration system and chemical have an expected life of 20 years at which time the chemical will have a resale value of $500 per kilogram, while the filtration system itself has no resale value.

Using a before-tax minimum attractive rate of return (MARR) of 10%, find the optimal amount of the chemical that should be purchased by the power plant.

Solution

Let X = number of kg of chemical purchased

Net Annual Cost (X) = (purchase cost of pollution abatement device)(A/P, 10%, 20)
+ (chemical purchase cost)(A/P, 10%, 20)
- (salvage value of chemical)(A/P, 10%, 20)
- (annual sale value of sulphur)

Net Annual Cost (X) = (5,000,000)(A/P, 10%, 20) + (1,000X)(A/P, 10%, 20)
- (500X)(A/F, 10%, 20) - (300) $(100\sqrt{X})$
$= 587.5 + 108.75X - 30{,}000\sqrt{X}$

To minimize cost differentiate Net Annual Cost (X) and set = 0

$$\frac{d\text{NAC(X)}}{dX} = 108.75 - \frac{30{,}000}{2\sqrt{X}} = 0$$

$$\frac{1}{2\sqrt{X}} = \frac{108.75}{30{,}000} \Rightarrow \sqrt{X} = \frac{30{,}000}{2(108.75)} = 137.931$$

$X = 137.931^2 = 19{,}024.97$

Check for max. or min. $\frac{d^2\text{NAC(X)}}{dX^2} = \frac{7{,}500}{\sqrt{X^3}} > 0$ For $X > 0$

9-45

The annual income from an apartment house is \$20,000. The annual expense is estimated to be \$2000. If the apartment house can be bought today for \$149,000, what is the breakeven resale price in ten years with 10% considered a suitable interest rate?

Solution

$P = (A_{INCOME} - A_{EXPENSES})(P/A, i\,\%, n) + F_{RE\text{-}SALE}(P/F, i\,\%, n)$
$149{,}000 = (20{,}000 - 2{,}000)(P/A, 10\%, 10) + F_{RE\text{-}SALE}\,(P/F, 10\%, 10)$
$149{,}000 = 18{,}000(6.145) + F_{RE\text{-}SALE}\,(.3855)$
$F_{RE\text{-}SALE} = \$99{,}584.95$

9-46

Oliver Douglas decides to install a fuel storage system for his farm that will save him an estimated 6.5 cents/gallon on his fuel cost. Initial cost of the system is $10,000 and the annual maintenance is a uniform gradient amount of $25. After a period of 10 years the estimated salvage is $3,000. If money is worth 12%, what is the breakeven quantity of fuel?

Solution

EAC =(10,000 - 3,000)(A/P, 12%, 10) + 3,000(.12) + 25(A/G, 12%, 10)
= $1,688.63

EAB = Gallons(.065) = $G(.065)

0 = -1,688.63 + G(.065)
G = 25,979 gallons

9-47

A land surveyor just starting in private practice needs a van to carry crew and equipment. He can lease a used van for $3,000 per year, paid at the beginning of each year, in which case maintenance is provided. Alternatively, he can buy a used van for $7,000 and pay for maintenance himself. He expects to keep the van three years at which time he could sell it for $1,500. What is the most he should pay for uniform annual maintenance to make it worthwhile buying the van instead of leasing it, if his MARR is 20%?

Solution

Lease:

EAC = 3,000(F/P, 20%, 1) = 3,000(1.20) = 3,600

Buy:

EAC = 7,000(A/P, 20%, 3) + M - 1,500(A/F, 20%, 3)
M = 3,600 - 2,910.85
= $ 689.15

9-48

The investment in a crane is expected to produce profit from its rental as shown below, over the next six years. Assume the salvage value is zero. Assuming 12% interest, what is the breakeven cost of the crane?

Year	Profit
1	$15,000
2	12,500
3	10,000
4	7,500
5	5,000
6	2,500

Solution

PW_{PROFIT} = 15,000(P/A, 12%, 6) - 2,500(P/G, 12%, 6) = \$39,340
$Cost_{BE}$ = \$39,340

9-49

ABC Manufacturing has a before-tax minimum attractive rate of return (MARR) of 12% on new investments. What uniform annual benefit would Investment B have to generate to make it preferable to Investment A?

Year	Investment A	Investment B
0	- $60,000	- $45,000
1 - 6	+15,000	?

Solution

NPW of A = - 60 + 15(P/A, 12%, 6) = 1.665

NPW of B $\geq$ 1.665 = - 45 + B(P/A, 12%, 6)
$\therefore$ B = 11,351

B > \$11,351 per year

9-50

Data for two drill presses under consideration by B&R Gears are listed below. Assuming an interest rate of 12%, what salvage value of press B will make the two alternatives equal?

	A	B
First cost	$30,000	$36,000
Annual maintenance	1,500	2,000
Salvage value	5,000	?
Useful life	6 years	6 years

Solution

EAC = P(A/P, i %, n) - S(A/F, i %, n) + Other Costs

DRILL PRESS A:

EAC = 30,000(A/P, 12%, 6) - 5,000(A/F, 12%, 6) + 1,500
= $8,180

DRILL PRESS B:

EAC = 36,000(A/P, 12%, 6) - SV(A/F, 12%, 6) + 2,000
= $10,755.20 - SV(.1232)

Setting the two EAC equal

8,180 = 10,755.20 - SV(.1232)
SV = $20,903

9-51

Dolphin Inc. trains mine seeking dolphins in a 5-mine tank. They are considering purchasing a new tank. A new tank costs $750,000 and realistic dummy mines cost $250,000. The new tank will allow the company to train 3 dolphins per year and will last 10 years costing $50,000 per year to maintain. How much must Dolphin Inc. receive (per dolphin) from the Navy in order to breakeven if the MARR equals 5%?

Solution

NPV = -Cost - Cost of Mines - Annual Maintenance(P/A, 5$, 10) + Income(P/A, 5%, 10)
= -750,000 - 250,000(5) - 50,000(P/A, 5%, 10) + 3(x)(P/A, 5%, 10)
x = $103,000

Chapter 10

Uncertainty in Future Events

10-1

Tee-to Green Golf Inc. is considering the purchase of new automated club assembly equipment. The industrial engineer for TGG thinks that she has determined the "best' choice. However she is uncertain how to evaluate the equipment because of questions concerning the actual annual savings and salvage value at the end of the expected life. The equipment will cost $500,000 and is expected to last eight years. Information concerning the savings and salvage value estimates and the projected probabilities is presented below:

	p = .20	p = .50	p = .25	p = .05
Savings per year	$65,000	$82,000	$90,000	$105,000
Salvage value	40,000	55,000	65,000	75,000

Determine the NPW if TGG's MARR is 6%.

Solution

E(Savings) = .2(65,000) + .5(82,000) + .25(90,000) + .05(105,000) = $81,750

E(Salvage) = .2(40,000) + .5(55,000) + .25(65,000) + .05(75,000) = $55,500

NPW = -500,000 + 81,750(P/A, 6%, 8) + 55,500(P/F, 6%, 8)
= $42,489

10-2

The two finalists in a tennis tournament are playing the championship. The winner will receive $60,000 and the runner-up $35,000. Determine the expected winnings for each participant if the players are considered to be evenly matched. What would the expected winnings be if one player were favored by 4 to 1 odds.

Solution

Evenly matched, both players expected winnings will be the same.

Winnings = .5(60,000) + .5(35,000) = $47,500

Assume player A is favored by 4 to 1 odds. The probability that A wins is then 4/5.

Player A's expected winnings = .8(60,000) + .2(35,000) = $55,000

Player B's expected winnings = .2(60,000) + .8(35,000) = $40,000

10-3

Consolidate Edison Power is evaluating the construction of a new electric generation facility. The two choices are a coal burning plant (CB) or a gaseous diffusion (GD) plant. The CB plant will cost $150 per megawatt to construct, and the GD plant will cost $300 per megawatt. Due to uncertainties concerning fuel availability and the impact of future air- and water- quality regulations, the useful life of each plant is unknown, but the following probability estimates have been made

	Probabilities	
Useful life (years)	CB plant	GD Plant
10	.10	.05
20	.50	.25
30	.30	.50
40	.10	.20

(a) Determine the expected life of each plant.
(b) Based on the ratio of construction cost per megawatt to expected life, which plant would you recommend ConEd build?

Solution

(a) Expected life

Coal burning = .10(10) + .50(20) + .30(30) + .10(40) = 24 years

Gaseous diffusion = .05(10) + .25(20) + .50(30) + .20(40) = 28.5 years

(b) Ratios

Coal burning = 150/24 = $6.25 per megawatt per year

Gaseous diffusion = 300/28.5 = $10.53 per megawatt per year

Recommend the coal burning plant

10-4

Palmer Potatoes Chips Inc. must purchase new potato peeling equipment for its Martin, Tennessee plant. The plant engineer has determined there are three possible set-ups that can be purchased. Relevant data are present below. All machines are expected to be used six years and PPC Inc,'s MARR is 10%. Which machine should be chosen?

Naked Peel			
First Cost	$45,000		
Annual Costs		p = .2	$3,000
		p = .7	4,500
		p = .1	5,500
Salvage Value		p = .7	$7,500
		p = .3	9,500
Skinner			
First Cost	$52,000		
Annual Costs		p = .4	$5,000
		p = .4	6,500
		p = .2	8,500
Salvage Value		p = .4	$5,500
		p = .3	7,500
		p = .3	8,500
Peel-O-Matic			
First Cost	$76,000		
Annual Costs		p = .3	$5,000
		p = .5	7,500
		p = .2	9,500
Salvage Value		p = .6	$8,500
		p = .4	9,000

Solution

Naked Peel

E(Annual costs) = .2(3000) + .7(4500) + .1(5500) = $4,300
E(Salvage value) = .7(7500) + .3(9500) = $8,100

NPW = -45,000 - 4300(P/A, 10%, 6) + 8100(P/F, 10%, 6) = -$59,154

Skinner

E(Annual costs) = $6,300
E(Salvage value) = $7,000

NPW = -$75,485

Peel-O-Matic

E(Annual costs) = \$7,150
E(Salvage value) = \$8,700

NPW = -\$102,227

Choose Naked Peel

10-5

Acme Insurance Company offers an insurance policy that pays \$1,000 for lost luggage on a cruise. Historically the company pays this amount in 1 out of every 200 policies it sells. What is the minimum amount Acme must charge for such a policy if they desire to make at least \$10 dollars per policy?

Solution

The probability that a loss occurs is $\frac{1}{200} = .005$

The expected loss to the company is therefore .005(1,000) = \$5

To make a profit of \$10 from each policy sold, they must charge \$15 per policy

10-6

A roulette wheel consists of 18 black slots, 18 red slots, and 2 green slots. If a \$100 bet is placed on black, what is the expected gain or loss? (A bet on black or red pays even money.)

Solution

The probability of black occurring = $\frac{18}{38}$

Expected value of the bet = $100\frac{18}{38} - 100\frac{20}{38} = \frac{100}{19} \approx \$5.26$

10-7

Crush Cola Company must purchase a bottle capping machine. Information concerning the machine and possible cash flows is presented below.

	p = .30	p = .50	p = .20
First Cost	$40,000	$40,000	$40,000
Annual Savings	2,000	3,500	5,000
Annual Costs	7,000	5,000	4,000
Actual Salvage Value	4,000	5,000	6,500

The machine is expected to have a useful life of 10 years and straight-line depreciation to a salvage value of $5,000 will be used. Crush pays taxes at the marginal rate of 34% and has a MARR of 12%. Determine the NPW of the machine.

Solution

$$\text{Depreciation} = \frac{40{,}000 - 5000}{10} = \$3{,}500$$

E(Saving/Costs) = -5,000(.30) – 1,500(.50) + 1,000(.20) = -$2,050

E(Salvage Value) = 4,000(.30) + 5,000(.50) + 6,500(.20) = $5,000

Year	BTCF	Depreciation	Taxable Income	Taxes	ATCF
0	-	-	-	-	-40,000
1-10	-2,050	3,500	-5,550	-1,887	-163
10	5,000	-			5,000

NPW = -40,000 - 163(P/A, 12%, 10) + 5,000(P/F, 12%, 10) = -$39,311

10-8

Krispy Kookies is considering the purchase of new dough mixing equipment. The NPW and estimated probabilities of the four possible outcomes are presented in the table below. Calculate the expected NPW of the equipment.

Outcome	NPW	Probability
1	$34,560	.15
2	38,760	.25
3	42,790	.40
4	52,330	.20

Solution

E(NPW) = .15(34,560) + .25(38,760) + .40(42,790) + .20(52,330) = \$42,456

10-9

Northeast Airlines is considering bidding for a new route to Asia. The route is expected to command a rather large price. There is uncertainty associated with all elements of the investment. The winner of the route will have use of it for five years before the bidding process is repeated. Information developed by Northeast concerning estimated cash flows and probabilities is summarized below. Based on a MARR of 8% determine the expected worth of the route.

Element	p = .15	p = .45	p = .30	p = .10
Bid Amount	11,500,000	22,250,000	27,500,000	38,250,000
Net Annual Income	3,000,000	5,700,000	7,000,000	9,600,000

Solution

NPW 1 = -11,500,000 + 3,000,000(P/A, 8%, 5) = \$479,000
NPW 2 = -22,250,000 + 5,700,000(P/A, 8%, 5) = \$510,100
NPW 3 = -27,500,000 + 7,000,000(P/A, 8%, 5) = \$451,000
NPW 4 = -38,250,000 + 9,600,000(P/A, 8%, 5) = \$ 82,800

E(NPW) = .15(479,000) + .45(510,100) + .30(451,000) + .10(82,800) = \$444,975

10-10

A new heat exchanger must be installed by CSI Inc. Alternative A has an initial cost of $33,400 and alternative B has an initial cost of $47,500. Both alternatives are expected to last ten years. The annual cost of operating the heat exchanger depends on ambient temperature in the plant and energy costs. The estimate of the cost and probabilities for each alternative is presented below. If CSI has a MARR of 8% and uses rate of return analysis for all capital decisions, which exchanger should they purchase?

	Annual Cost	Probability
Alternative A	$4,500	.10
	7,000	.60
	8,000	.25
	9,250	.05
Alternative B	$4,000	.20
	5,275	.60
	6,450	.15
	8,500	.05

Solution

Alternative A

E(Annual Cost) = .10(4,500) + .60(7,000) + .25(8,000) + .05(9,250) = $7112.50

Alternative B

E(Annual Cost) = .20(4,000) + .60(5,275) + .15(6,450) + .05(8,500) = $5,357.50

Incremental Analysis is required

<u>B-A</u>

NPW = 0 at IRR

0 = (-47,500 -(-33,400)) + (-5,375.50 - (-7112.50))(P/A, i%, 10)
0 = -14,100 + 1737(P/A, i%, 10)

(P/A, i%, 10) = 8.12

i = 4% P/A = 8.111

IRR ≈ 4%

CSI should purchase the least expensive alternative, A

10-11

The probability that a machine will last a certain number of years is given in the following table.

Years of Life	Probability of Obtaining Life
10	0.15
11	0.20
12	0.25
13	0.20
14	0.15
15	0.05

What is the expected life of the machine?

Solution

$$\begin{aligned}\text{Expected value} &= 10(0.15) + 11(0.20) + 12(0.25) + 13(0.20) + 14(0.15) + 15(0.05)\\ &= 12.15 \text{ years}\end{aligned}$$

Chapter 11

Depreciation

11-1

Some seed cleaning equipment was purchased in 1995 for $8,500 and is depreciated by the double declining balance (DDB) method for an expected life of 12 years. What is the book value of the equipment at the end of 2000? Original salvage value was estimated to be $2,500 at the end of 12 years.

Solution

$$\text{Book Value} = P(1 - \tfrac{2}{N})^n$$

$$= 8{,}500(1 - \tfrac{2}{12})^6 = \$2{,}846.63$$

This can be checked by doing the year-by-year computations:

YEAR	DDB	
1995	(8,500-0)	= $1,416.67
1996	(8,500-1,416.67) =	1,180.56
1997	(8,500-2,597.23) =	983.80
1998	(8,500-3,581.03) =	819.83
1999	(8,500-4,400.86) =	683.19
2000	(8,500-5,084.05) =	569.32

Book Value = 8,500 – 5,653.37 = $2,846.63

11-2

Suds-n-Dogs just purchased new automated wiener handling equipment for $12,000. The salvage value of the equipment is anticipated to be $1,200 at the end of its five year life. Using MACRS, determine the depreciation schedule.

Solution

Three year class is determined.

Year		Depreciation
1	12,000(.3333)	$3,999.60
2	12,000(.4445)	5,334.00
3	12,000(.1481)	1,777.20
4	12,000(.0741)	889.20

11-3

An asset will cost $1,750 when purchased this year. It is further expected to have a salvage value of $250 at the end of its five year depreciable life. Calculate complete depreciation schedules giving the depreciation charge, D(n), and end-of-year book value, B(n), for straight line (SL), sum of the years digits (SOYD), double declining balance (DDB), and modified accelerated cost recovery (MACRS) depreciation methods. Assume a MACRS recovery period of 5 years.

Solution

	SL		SOYD		DDB		MACRS	
n	D(n)	B(n)	D(n)	B(n)	D(n)	B(n)	D(n)	B(n)
0		1,750		1,750		1,750		1,750.00
1	300	1,450	500	1,250	700	1,050	350.00	1,400.00
2	300	1,150	400	850	420	630	560.00	840.00
3	300	850	300	550	252	378	336.00	504.00
4	300	550	200	350	128	250	201.60	302.40
5	300	250	100	250	0	250	201.60	100.80
6							100.80	0.00

11-4

Your company is considering the purchase of a second-hand scanning microscope at a cost of $10,500, with an estimated salvage value of $500 and a projected useful life of four years. Determine the straight line (SL), sum of years digits (SOYD), and double declining balance (DDB) depreciation schedules.

Solution

Year	SL	SOYD	DDB
1	2,500	4,000	5,250.00
2	2,500	3,000	2,625.00
3	2,500	2,000	1,312.50
4	2,500	1,000	656.25

11-5

A piece of machinery costs $5,000 and has an anticipated $ 1,000 resale value at the end of its five year useful life. Compute the depreciation schedule for the machinery by the sum-of-years-digits method.

Solution

$$\text{Sum - of - years - digits} = \tfrac{n}{2}(n + 1) = \tfrac{5}{2}(6) = 15$$

1st year depreciation = $\frac{5}{15}$(5,000 - 1,000) = $1,333
2nd year depreciation = $\frac{4}{15}$(5,000 - 1,000) = 1,067
3rd year depreciation = $\frac{3}{15}$(5,000 - 1,000) = 800
4th year depreciation = $\frac{2}{15}$(5,000 - 1,000) = 533
5th year depreciation = $\frac{1}{15}$(5,000 - 1,000) = 267

11-6

A new machine costs $12,000 and has a $1200 salvage value after using it for eight years. Prepare a year-by-year depreciation schedule by the double declining balance (DDB) method.

Solution

$$\text{DDB Deprecation} = \tfrac{2}{N}(P - \Sigma D)$$

Year	1	2	3	4	5	6	7	8*	Total
Deprecation	3,000	2,250	1,688	1,266	949	712	534	401	$10,800

*Book value cannot go below declared salvage value. Therefore the full value of year eight's depreciation cannot be taken.

11-7

To meet increased sales, a large dairy is planning to purchase 10 new delivery trucks. Each truck will cost $18,000. Compute the depreciation schedule for each truck, using the modified accelerated cost recovery system (MACRS) method, if the recovery period is 5 years.

Solution

Year		Depreciation
1	18K(.20)	$3,600.00
2	18K(.32)	5,760.00
3	18K(.192)	3,456.00
4	18K(.1152)	2,073.60
5	18K(.1152)	2,073.60
6	18K(.0576)	1,036.80

11-8

The XYZ Research Company purchased a total of \$215,000 worth of assets in the 2003. Determine the MACRS depreciation schedule using the maximum allowable 179 expenses.

Solution

Maximum 179 = \$25,000

Assets purchased = \$215,000, the maximum allowable 179 = 25,000 – (215,000 – 200,000)
= 15,000

Research equipment is five year recovery class

Basis = 215,000 - 15,000
= 210,000

Year		Depreciation
1	205K(.20)	\$41,000
2	205K(.32)	65,600
3	205K(.192)	39,360
4	205K(.1152)	23,616
5	205K(.1152)	23,616
6	205K(.0576)	11,808

11-9

A used piece of depreciable property was bought for \$20,000. If it has a useful life of 10 years and a salvage value of \$5,000, how much will it be depreciated in the 9th year, using the 150% declining balance schedule?

Solution

$$\text{Depreciation} = \frac{1.5P}{N}\left(1-\frac{1.5}{N}\right)^{n-1} = \frac{1.5(20{,}000)}{10}\left(1-\frac{1.5}{10}\right)^{9-1} = \$817.50$$

Check BV at end of 8th year

$$BV = P\left(1-\frac{1.5}{N}\right)^{n} = 20{,}000\left(1-\frac{1.5}{10}\right)^{8} = \$5{,}449.80$$

Because the salvage value is \$5,000 in the 9th year you can only depreciate \$449.80 (5,449.80 - 5,000). The \$817.50 would have brought the book value below the salvage value of \$5,000.

11-10

A front-end loader cost $70,000 and has an estimated salvage value of $10,000 at the end of 5 years useful life. Compute the depreciation schedule, and book value, to the end of the useful life of the tractor using MACRS depreciation.

Solution

Five year recovery period is determined.

Year		Depreciation	Book Value
1	70K(.20)	$14,000	70,000 - 14,000 = $56,000
2	70K(.32)	22,400	56,000 - 22,400 = 33,600
3	70K(.192)	13,440	33,600 - 13,440 = 20,160
4	70K(.1152)	8,064	20,160 - 8,064 = 12,096
5	70K(.1152)	8,064	12,096 - 8,064 = 4,032
6	70K(.0576)	4,032	4,032 - 4,032 = 0

11-11

A machine costs $5000 and has an estimated salvage value of $1000 at the end of 5 years useful life. Compute the depreciation schedule for the machine by

(a) Straight line (SL)
(b) Double declining balance (DDB)
(c) Sum of years digits (SOYD)

Solution

(a) $\text{SL} = \frac{\text{P-S}}{\text{N}} = \frac{5{,}000-1{,}000}{5} = \$800/\text{year}$

(b) $\text{DDB} = \frac{2}{\text{N}}\left[\text{P} - \Sigma \text{D}_\text{c}(t)\right]$

			Depreciation
D_c Year 1 =	.4[5,000 - 0]	= $2,000	$2,000
D_c Year 2 =	.4[5,000 - 2,000]	= 1,200	1,200
D_c Year 3 =	.4[5,000 - 3,200]	= 720	720
D_c Year 4 =	.4[5,000 - 3,920]	= 432 →	80
D_c Year 5 =	.4[5,000 - 4,352]	= 259.20→	0
		$4,611.20	$4,000

(c) $\text{SOYD} = \frac{N}{2}(N+1) = \frac{5}{2}(6) = 15$

		Depreciation
D_c Year 1 =	$\frac{5}{15}(5{,}000 - 1{,}000)$	= \$1,333
D_c Year 2 =	$\frac{4}{15}(5{,}000 - 1{,}000)$	= 1,067
D_c Year 3 =	$\frac{3}{15}(5{,}000 - 1{,}000)$	= 800
D_c Year 4 =	$\frac{2}{15}(5{,}000 - 1{,}000)$	= 533
D_c Year 5 =	$\frac{1}{15}(5{,}000 - 1{,}000)$	= 267

11-12

A lumber company purchased a tract of timber for \$70,000. The value of the 25,000 trees on the tract was established to be \$50,000. The value of the land was established to be \$20,000. In the first year of operation, the lumber company cut down 5000 trees. What was the depletion allowance for the year?

Solution

For standing timber only cost depletion (not percentage depletion) is permissible. Five thousand of the trees were harvested therefore 5,000/25,000 = 0.20 of the tract was depleted. Land is not considered depletable, only the timber, which is valued at a total of \$50,000.

Therefore, the first year's depletion allowance would be = 0.20(\$50,000) = \$10,000.

11-13

A machine was purchased two years ago for \$50,000 and had a depreciable life of five years and no expected salvage value. The owner is considering an offer to sell the machine for \$25,000. For each of the depreciation methods listed, fill in the table below to determine the deprecation for year 2, and the book value at the end of year 2. (Assume a five year MARCS recovery period.)

	Depreciation For Year 2	End of Year 2 Book Value
Sum-Of-Years-Digits (SOYD)		
Straight Line (SL)		
Double Declining Balance (DDB)		
Modified Accelerated Cost Recovery System (MACRS)		

Solution

	Depreciation For Year 2	End of Year 2 Book Value
Sum-Of-Years-Digits (SOYD)	\$13,333	\$20,000
Straight Line (SL)	10,000	30,000
Double Declining Balance (DDB)	12,000	18,000
Modified Accelerated Cost Recovery System (MACRS)	16,000	24,000

SOYD

$\text{SOYD} = \frac{N}{2}(N+1) = \frac{5}{2}(6) = 15$

depreciation year 1 = $\frac{5}{15}(50{,}000-0)$ = \$16,666

depreciation year 2 = $\frac{4}{15}(50{,}000-0)$ = \$13,333

cumulative depreciation = 16,666 + 13,333 = \$30,000

book value = P - cumulative depreciation = 50,000 - 30,000 = \$20,000

SL

depreciation = (P - S)/N = (50,000 - 0)/5 = \$10,000 per year

cumulative depreciation = 2 x 10,000 = \$20,000

book value = 50,000 - 20,000 = \$30,000

DDB:

depreciation year 1 = $\frac{2}{N}$ (Book Value) = $\frac{2}{5}$ (50,000) = \$20,000

depreciation year 2 = $\frac{2}{5}$ (50,000 - 20,000) = \$12,000

cumulative depreciation = 20,000 + 12,000 = \$32,000

book value = 50,000 - 32,000 = \$18,000

MACRS:

depreciation year 1 = (20%)(50,000) = \$10,000

depreciation year 2 = (32%)(50,000) = \$16,000

cumulative depreciation = 10,000 + 16,000 = \$26,000

book value = 50,000 - 26,000 = \$24,000

11-14

In the production of beer, a final filtration is given by the use of "Kieselguhr" or diatomaceous earth, which is composed of the fossil remains of minute aquatic algae, a few microns in diameter and composed of pure silica. A company has purchased a property for $840,000 that contains an estimated 60,000 tons. Compute the depreciation charges for the first three years, if a production (or extraction) of 3000 tons, 5000 tons, and 6000 tons are planned for years 1, 2, and 3, respectively. Use the cost-depletion methods, assuming no salvage value for the property.

Solution

Total diatomaceous earth in property = 60,000 tons
Cost of property = $480,000

Then, $\frac{\text{depletion allowance}}{\text{tons extracted}} = \frac{\$840{,}000}{60{,}000 \text{ tons}} = \$14 / \text{ton}$

Year	Tons Extracted	Depreciation Charge
1	3,000	3,000 x 14 = $42,000
2	4,000	4,000 x 14 = 56,000
3	5,000	5,000 x 14 = 70,000

11-15

A pump cost $1,000 and has a salvage value of $100 after a life of five years. Using the double declining balance depreciation method, determine:

(a) The depreciation in the first year.
(b) The book value after five years.
(c) The book value after five years if the salvage was only $50.

Solution

a) Rate = $\frac{200\%}{5} = 40\% = .4$
$1{,}000(.4) = \$400$

b) Book Value = $P(1 - \frac{2}{N})^n$
B.V. = maximum of $\{\text{s.v.}; 1{,}000(1-.4)^5\}$ = maximum of $\{100, 77.76\}$ = $100

c) B.V. = maximum of $\{\text{s.v.}; 1{,}000(1-.4)^5\}$ = maximum of $\{50, 77.76\}$ = $77.76

11-16

Adventure Airlines recently purchased a new baggage crusher for $50,000. It is expected to last for 14 years and have an estimated salvage value of $8,000. Determine the depreciation charge on the crusher for the third year of its life and the book value at the end of 8 years, using sum-of-digits depreciation.

Solution

SOYD depreciation for 3rd year

$$\text{Sum of Years digits} = \frac{n}{2}(n+1) = \frac{14}{2}(14+1) = 105$$

$$3^{rd}\text{ year depreciation} = \frac{\text{remaining life at beginning of year}}{\sum \text{years digits}}(P - S)$$

$$= \frac{12}{105}(50{,}000 - 8{,}000) = \$4{,}800$$

Book Value at end of 8 years

$$\sum 8 \text{ years of depreciation} = \frac{14+13+12+11+10+9+8+7}{105}(50{,}000 - 8{,}000)$$

$$= \frac{84}{105}(42{,}000) = \$33{,}600$$

Book Value = Cost - Depreciation to date = 50,000 - 33,600 = $16,400

11-17

A small wood chipping company purchases a new chipping machine for $150,000. The total cost of assets purchased for the tax year 2001 is $210,000. If the company chooses to use their entire 179 expense on the wood chipper, determine the MACRS depreciation schedule. (ADR of the chipper is 12 years.)

Solution

Maximum 179 = $24,000 (for tax year 2001)

Assets purchased = $210,000, the maximum allowable 179 = 24,000 – (210,000 – 200,000)
= 14,000

ADR 12 years → 7 year recovery period

Basis = 150,000 - 14,000
= 136,000

Year		Depreciation
1	136K(.1429)	$19,434.40
2	136K(.2449)	33,306.40
3	136K(.1749)	23,786.40
4	136K(.1249)	16,986.40
5	136K(.0893)	12,144.80
6	136K(.0892)	12,131.20
7	136K(.0893)	12,144.80
8	136K(.0446)	6,065.60

Chapter 12

Income Taxes

12-1

A tool costing $300 has no salvage value. Its resultant cash flow before tax is shown below.

Year	BTCF	SOYD Depreciation	Taxable Income	Taxes	ATCF
0	-$300				
1	+100				
2	+150				
3	+200				

The tool must be depreciated over 3 years according to the sum of the years digits method. The tax rate is 35%.

(a) Fill in the four columns in the table.
(b) What is the internal rate of return after tax?

Solution

a)

Year	BTCF	SOYD Depreciation	Taxable Income	Taxes	ATCF
0	-$300				-300.00
1	+100	150	-50	-17.50	+117.50
2	+150	100	+50	+17.50	+132.50
3	+200	50	+150	+52.50	+147.50

b) NPW = -300 + 117.50(P/A, i%, 3) + 15(P/G, i%, 3) = 0

By trial and error.

Try i = 12% NPW = $15.55
Try i = 15% NPW = -$ 0.68

IRR = i ≈ 15% (By interpolation i = 14.87%)

12-2

A company, whose earnings put them in the 35% marginal tax bracket, is considering purchasing a piece of equipment for $25,000. The equipment will be depreciated using the straight line method over a 4 year useful life to a salvage value of $5,000. It is estimated that the equipment will increase the company's earnings by $8,000 for each of the 4 years. Should the equipment be purchased? Use an interest rate of 10%.

Solution

Year	BTCF	Depreciation	T.I	Taxes	ATCF
0	-25,000				-25,000
1	8,000	5,000	3,000	1,050	6,950
2	8,000	5,000	3,000	1,050	6,950
3	8,000	5,000	3,000	1,050	6,950
4	8,000	5,000	3,000	1,050	6,950
	5,000				5,000

StL Depreciation = ¼(25,000 - 5,000) = $5,000

Find net present worth: if > 0 than good deal

NPV = -25,000 + 6,950(P/A, 10%,4) + 5,000(P/F, 10%, 4) = $446.50

Purchase equipment

12-3

By purchasing a truck for $30,000, a large profitable company in the 34% income tax bracket was able to save $8000 during year 1, $7000 during year 2, $6000 during year 3, $5000 during year 4, and $4000 during year 5. The company depreciated the truck using the sum-of-years-digits depreciation method over its four year depreciable life, while assuming a zero salvage value for depreciation purpose. The company wants to sell the truck at the end of year 5. What resale value will yield a 12% after-tax rate of return for the company?

Solution

$$\text{SOYD} = \frac{4(5)}{2} = 10$$

Year	SOYD Depreciation
1	4/10(30,000 - 0) = 12,000
2	3/10(30,000 - 0) = 9,000
3	2/10(30,000 - 0) = 6,000
4	1/10(30,000 - 0) = 3,000

ATCF Calculations:

Year	BTCF	Depreciation	T.I.	Taxes	ATCF
0	-30,000				-30,000
1	8,000	12,000	-4,000	-1,360	9,360
2	7,000	9,000	-2,000	-680	7,680
3	6,000	6,000	-0,000	0	6,000
4	5,000	3,000	2,000	+680	4,320
5	4,000	--	4,000	+1,360	2,640
	Resale	--	Resale	.34Resale	.66Resale

Solve for Resale (R):

30,000 = 9,360(P/A, 12%, 5) - 1,680(P/G, 12%, 5) + .66R(P/F, 12%, 5)
30,000 = 22,995.84 + .374484R

R = $18,703.50

12-4

A company has purchased a major piece of equipment, which has a useful life of 20 years. An analyst is trying to decide on a maintenance program and has narrowed the alternatives to two. Alternative A is to perform $1000 of major maintenance every year. Alternative B is to perform $5000 of major maintenance only every fourth year. In either case, maintenance will be performed during the last year so that the equipment can be sold for an established $10,000. If the MARR is 18%, which maintenance plan should be chosen?

The analyst computed the solution as:
Equivalent Annual $Cost_A$ = $1000
Equivalent Annual $Cost_B$ = $5000 (A/F, 18%, 4) = $958.50
Therefore choose Alternative B.
Is it possible the decision would change if income taxes were considered? Why or Why not?

Solution

No. Both cash flow, which distinguish between the alternative, would be reduced by the same percentage (i.e. taxes) so $EAC_A > EAC_B$ would still be true. If we assume a 45% tax rate, for example, the computations as follows:

Alternative A	Year	BTCF	T.I.	Taxes	ATCF
	1-4	-1,000	-1,000	-450*	-550

$EAC_A = 550$

Alternative B	Year	BTCF	T.I.	Taxes	ATCF
	1-3	0	0	0	0
	4	-5,000	-5,000	-2,250*	-2,750

$EAC_B = 2{,}750(A/F, 18\%, 4) = 527.20$

*where + sign indicates a decrease in income taxes.

12-5

A large company must build a bridge to have access to land for expansion of its manufacturing plant. The bridge could be fabricated of normal steel for an initial cost of $30,000 and should last 15 years. Maintenance will cost $1000/year. If more corrosion resistant steel were used, the annual maintenance cost would be only $100/year, although the life would be the same. In 15 years there will be no salvage value for either bridge. The company pays a combined state and federal income tax rate of 48% and uses straight line depreciation. If the minimum attractive after tax rate of return is 12%, what is the maximum amount that should be spent on the corrosion resistant bridge?

Solution

Steel

Year	BTCF	Depreciation	T.I.	Taxes	ATCF
0	-30,000				-30,000
1-15	-1,000	2,000	-3,000	-1,440	+440

Corrosion Resistant Steel

Year	BTCF	Depreciation	T.I.	Taxes	ATCF
0	-P				-P
1-15	-100	P/15	-100 - P/15	(-48 - .032P)	-52 + .032P

$NPW_A = NPW_B$ for breakeven

$$440(P/A, 12\%, 15) - 30{,}000 = (-52 + .032P)(P/A, 12\%, 15) - P$$
$$-27{,}003 = -354 + .0281P - P$$
$$P = \$34{,}078$$

12-6

An unmarried, recent engineering graduate earned $52,000 during the previous tax year. The engineer claims one $3000 exemption and establishes itemized deductions of $5,200. If single people are automatically allowed $4,700 of deductions, what is the student's federal income tax?

Tax Rate:

Tax Rate	Income
10%	< $6,000
15%	$6,000 to $27,950
27%	$27,950 to $67,700

Solution

Adjusted Gross Income	$52,000
Exemption	- 3,000
Itemized deductions (5,200 - 4,700)	- 500
Taxable Income	$48,500

Tax = 6,000(.10) + (27,950 - 6,000)(.15) + (48,500 - 27,950)(.27)
= $9,441

12-7

An asset with five year MACRS* life will be purchased for $10,000. It will produce net annual benefits of $2000 per year for six years, after which it will have a net salvage value of zero and will be retired. The company's incremental tax is 34%. Calculate the after tax cash flows.

*The annual percentages to use are 20, 32, 19.20, 11.52, 11.52, and 5.76 for years 1 through 6.

Solution

Year	BTCF	Depreciation	T.I.	Taxes	ATCF
0	-10,000				-10,000.00
1	2,000	2,000	0	0	2,000.00
2	2,000	3,200	-1,200	-408.00	2,408.00
3	2,000	1,920	80	+27.20	1,972.80
4	2,000	1,152	848	+288.32	1,711.68
5	2,000	1,152	848	+288.32	1,711.68
6	2,000	576	1,424	+484.16	1,515.84

12-8

A state tax of 10% is deductible from the income taxed by the federal government (Internal Revenue Service). The federal tax is 34%. What is the combined effective tax rate?

Solution

$$\begin{aligned}\text{Effective rate} &= S + (1 - S)F \\ &= .1 + (1 - .1).34 \\ &= .1 + (.9).34 \\ &= .1 + .306 \\ &= .406 = 40.6\%\end{aligned}$$

12-9

For engineering economic analysis a corporation uses an incremental state income tax rate of 7.4% and an incremental federal rate of 34%. Calculate the effective tax rate.

Solution

$$\begin{aligned}\text{Effective rate} &= S + (1 - S)F \\ &= .074 + (1 - .074)(.34) \\ &= .074 + (.926).34 \\ &= .074 + .3148 \\ &= .3888 = 38.88\%\end{aligned}$$

12-10

A company bought an asset at the beginning of 2001 for $100,000. The company now has an offer to sell the asset for $60,000 at the end of 2002. For each of the depreciation methods shown below determine the capital loss or recaptured depreciation that would be realized for 2002.

Depreciation Method	Depreciable Life	Salvage Value*	Recaptured Depreciation	Capital Loss
SL	10 years	$ 1,000		
SOYD	5	25,000		
DDB	4	0		
150%DB	15	0		

*This was assumed for depreciation purposes

Solution

Depreciation Method	Depreciable Life	Salvage Value*	Recaptured Depreciation	Capital Loss
SL	10 years	$ 1,000		$20,200
SOYD	5	25,000	$ 5,000	
DDB	4	0	35,000	
150%DB	15	0		21,000

SL:

Depreciation $= \frac{1}{10}(100{,}000 - 1{,}000) = 9{,}900$

Book Value $= 100{,}000 - 2(9{,}900) = 80{,}200$

Capital Loss $= 80{,}200 - 60{,}000 = \$20{,}200$

SOYD:

Depreciation (Year 1 + Year 2) $= \frac{5+4}{15}(100{,}000 - 25{,}000) = 45{,}000$

Book Value $= 100{,}000 - 45{,}000 = 55{,}000$

Recaptured Depreciation $= 60{,}000 - 55{,}000 = \$5{,}000$

DDB:

Depreciation Year 1 $= \frac{2}{4}(100{,}000) = 50{,}000$

Depreciation Year 2 $= \frac{2}{4}(100{,}000 - 50{,}000) = 25{,}000$

Book Value $= 100{,}000 - 75{,}000 = 25{,}0000$

Recaptured Depreciation $= 60{,}000 - 25{,}000 = \$35{,}000$

150%DB:

Depreciation Year 1 $= \frac{1.5}{15}(100{,}000) = 10{,}000$

Depreciation Year 2 $= \frac{1.5}{15}(100{,}000 - 10{,}000) = 9{,}000$

Book Value $= 100{,}000 - 19{,}000 = 81{,}000$

Capital Loss $= 81{,}000 - 60{,}000 = \$35{,}000$

12-11

A corporation's tax rate is 34%. An outlay of $35,000 is being considered for a new asset. Estimated annual receipts are $20,000 and annual disbursements $10,000. The useful life of the asset is 5 years and it has no salvage value.

(a) What is the prospective internal rate of return (IRR) before income tax?

(b) What is the prospective internal rate of return (IRR) after taxes, assuming straight line depreciation for writing off the asset for tax purpose?

Solution

$$D_C(SL) = \frac{1}{N}(P - S) = \frac{25{,}000 - 0}{5} = 7{,}000/\text{year.}$$

Year	BTCF	Depreciation	T.I.	Taxes	ATCF
0	-35,000				-35,000
1	10,000	7,000	3,000	1,020	8,980
2	10,000	7,000	3,000	1,020	8,980
3	10,000	7,000	3,000	1,020	8,980
4	10,000	7,000	3,000	1,020	8,980
5	10,000	7,000	3,000	1,020	8,980

a) $ROR_{BEFORE\ TAX}$

$PW_B = PW_C$

A(P/A, i%, n) = First Cost

$(P/A, i\%, 5) = \frac{35,000}{10,000} = 3.500$

12% = 3.605
P/A = 3.500
15% = 3.352

By interpolation:

IRR = 13.25%

b) $ROR_{AFTER\ TAX}$

A(P/A, i%, n) = First Cost

$(P/A, i\%, 5) = \frac{35,000}{8,980} = 3.898$

8% = 3.993
P/A = 3.898
9% = 3.890

By interpolation:

IRR = 8.92%

12-12

A large and profitable company, in the 34% income tax bracket, is considering the purchase of a new piece of machinery that will yield benefits of $10,00 for year 1, $15,000 for year 2, $20,000 for year 3, $20,000 for year 4, and $20,000 for year 5.

The machinery is to be depreciated using the modified accelerated cost recovery system (MACRS) with a three year recovery period. The percentages are 33.33, 44.45, 14.81, 8.41, respectively, for years 1, 2, 3, and 4. The company believes the machinery can be sold at the end of five years of use for 25% of the original purchase price.

What is the maximum purchase cost the company can pay if it requires a 12% after-tax rate of return?

Solution

Year	BTCF	Depreciation	T.I.	Taxes	ATCF
0	-P				-P
1	10,000	.3333P	10,000 - .3333P	3,400 + .1133P	6,600 + .1133P
2	15,000	.4445P	15,000 - .4445P	5,100 + .1511P	9,900 + .1511P
3	20,000	.1481P	20,000 - .1481P	6,800 + .0504P	13,200 + .0504P
4	20,000	.0841P	20,000 - .0841P	6,800 + .0286P	13,200 + .0286P
5	20,000	-	20,000	6,800	13,200
	.25P		.25P	.085P	.165P

P = 6,600(P/A 12%, 3) + 3,300(P/G, 12%, 3) + 13,200(P/A, 12%, 2)(P/F, 12%, 3)
+ .1133P(P/F, 12%, 1) + .1511P(P/F, 12%, 2) + .0504P(P/F, 12%, 3)
+ .0285P(P/F, 12%, 4) + .165P(P/F, 12%, 5)
P = $61,926.52

12-13

A manufacturing firm purchases a machine in January for 100,000. The machine has an estimated useful life of 5 years, with an estimated salvage value of $20,000. The use of the machine should generate $40,000 before-tax profit each year over its 5-year useful life. The firm pays combined taxes at the 40% and uses sum of the years digits depreciation.

PART 1 Complete the following table.

Year	BTCF	Depreciation	T.I.	Taxes	ATCF
0					
1					
2					
3					
4					
5					

PART 2 Does the sum of digits depreciation represent a cash flow?

PART 3. Calculate the before-tax rate of return and the after-tax rate of return.

Solution

PART 1:

Before Tax Cash Flow --

In January you must pay $100,000. At the end of the first year, and in each subsequent year you realize a cash income of $40,000. This income less depreciation allowance is taxable. In year 5 you also realize $20,000 from salvage of the equipment. This amount is not taxable as it represents a capital expense that was never allocated as depreciation.

Sum of Years Digits Depreciation --

$$\text{Year } 1 = \tfrac{5}{15}(100{,}000 - 20{,}000) = \$26{,}667$$

$$\text{Year } 2 = \tfrac{4}{15}(80{,}000) = \$21{,}333$$

$$\text{Year } 3 = \tfrac{3}{15}(80{,}000) = \$16{,}000$$

$$\text{Year } 4 = \tfrac{2}{15}(80{,}000) = \$10{,}667$$

$$\text{Year } 5 = \tfrac{1}{15}(80{,}000) = \$\ 5{,}333$$

Year	BTCF	Depreciation	T.I.	Taxes	ATCF
0	-100,000				-100,000
1	40,000	26,667	13,333	-5,333	34,667
2	40,000	21,333	18,667	-7,467	32,533
3	40,000	16,000	24,000	-9,600	30,400
4	40,000	10,667	29,333	-11,733	28,267
5	40,000	5,333	34,667	-13,867	26,133
5	20,000	--	--	--	20,000

PART 2:

The sum of digits depreciation is a bookkeeping allocation of capital expense for purposes of computing taxable income. In itself it dose not represent a cash flow.

PART 3:

Before Tax Rate of Return --

NPW = 0 at IRR

0 = -100,000 + 40,000(P/A, i%, 5) + 20,000(P/F, i%, 5)
Try 30%* NPW = +2,826
Try 35% NPW = -6,740

By interpolation:

IRR = 31.5%

After Tax Rate of Return --

0 = -100,000 + 34,667(P/A, i%, n) - 2,133(P/G, i%, n) + 20,000(P/F, i%, n)
Try 20%* NPW = +1,263
Try 25% NPW = -9,193

By interpolation:
IRR = 20.60%

*Caution should be exercised when interpolating to find an interest rate. Linear interpolation is being imposed on a curvilinear function. Therefore the solution is approximate. A maximum range over which to interpolate and achieve generally good results is usually considered to be three percentage points. Often times the interest tables you have at your disposal will force you to use a larger range. This is the case in this problem.

12-14

PARC, a large profitable firm, has an opportunity to expand one of its production facilities at a cost of $375,000. The equipment is expected to have an economic life of 10 years and to have a resale value of $25,000 after 10 years of use. If the expansion is undertaken, PARC expects that their income will increase by $60,000 for year 1, and then increase by $5000 each year through year 10 ($65,000 for year 2, $70,000 for year 3, etc.). The annual operating cost is expected to be $5000 for the first year and to increase by 250 per year ($5250 year 2, $5500 for year 3, etc.). If the equipment is purchased, PARC will depreciate it using straight line to a zero salvage value at the end of year 8 for tax purposes. Since PARC is a "large and profitable" firm their tax rate is 34%.

If PARC's minimum attractive rate of return (MARR) is 15%, should they undertake this expansion?

Solution

Year	Income	Cost	BTCF
1	60,000	5,000	55,000
2	65,000	5,250	59,750
3	70,000	5,500	64,500
4	75,000	5,750	69,250
5	80,000	6,000	74,000
6	85,000	6,250	78,750
7	90,000	6,500	83,500
8	95,000	6,750	88,250
9	100,000	7,000	93,000
10	105,000	7,250	97,750

Year	BTCF	Depreciation	T.I.	Taxes	ATCF
0	-375,000				-375,000.00
1	55,000	46,875	8,125	2,762.50	52,237.50
2	59,750	46,875	12,875	4,377.50	55,372.50
3	64,500	46,875	17,625	5,992.50	58,507.50
4	69,250	46,875	22,375	7,607.50	61,642.50
5	74,000	46,875	27,125	9,222.50	64,777.50
6	78,750	46,875	31,875	10,837.50	67,912.50
7	83,500	46,875	36,625	12,452.50	71,047.50
8	88,250	46,875	41,375	14,067.50	74,182.50
9	93,000	--	93,000	31,620.00	61,380.00
10	97,750	--	97,750	31,620.00	61,380.00
10	25,000	--	25,000	8,500.00	16,500.00

NPW = -375,000 + [52,237.50(P/A, 15%, 8) + 3,135(P/G, 15%, 8)] + 61,380(P/F, 15%, 9) + 77,880(P/F, 15%, 10)

NPW = -$64,780.13

PARC should not undertake the expansion.

12-15

A young man bought a one-year savings certificate for $10,000, which pays 15%. Before he buys the certificate he has a taxable income that puts him in the 27% incremental income tax rate, what is his after tax rate of return on this investment?

Solution

After tax ROR = (1 - tax rate)(before tax IRR) = (1 - .27)(.15) = 10.95%

12-16

A project can be summarized by the data given in the table below. The company uses straight line depreciation, pays an incremental income tax rate of 30% and requires an after-tax rate of return of 12%.

First Cost	$75,000
Annual Revenues	26,000
Annual Costs	13,500
Salvage Value	15,000
Useful Life	30 years

(a) Using Net Present Worth, determine whether the project should be undertaken.
(b) If the company used Sum-Of-Years-Digits depreciation, is it possible the decision would change? (No computations needed.)

Solution

a)

Year	BTCF	Depreciation	T.I.	Taxes	ATCF
0	-75,000				-75,000
1-30	12,500	2,000	10,500	3,150	9,350
30	15,000	--	--	--	15,000

NPW = -75,000 + 9,350(P/A, 12%, 30) + 15,000(P/F, 12%, 30) = $815

Yes, take project since NPW > 0

b) No. Although total depreciation is the same, SOYD is larger in the early years when it is worth more. Therefore the NPW would increase with SOYD making the project even more desirable.

12-17

A corporation expects to receive $32,000 each year for 15 years if a particular project is undertaken. There will be an initial investment of $150,000. The expenses associated with the project are expected to be $7530 per year. Assume straight line depreciation, a 15-year useful life and no salvage value. Use a combined state and federal 48% income tax rate and determine the project after-tax rate of return.

Solution

Year	BTCF	Depreciation	T.I.	Taxes	ATCF
0	-150,000				-150,000
1-30	24,470	10,000	14,470	6,946	17,524

Take the ATCF and compute the rate of return at which PW_{COSTS} equals $PW_{BENEFITS}$.

$$150{,}000 = 17{,}524(P/A, i\%, 15)$$

$$(P/A, i\%, 15) = \frac{150{,}000}{17{,}524} = 8.559$$

From interest tables, i = 8%

12-18

A project will require the investment of $108,000 in capital equipment (SOYD with a depreciable life of eight years and zero salvage value) and $25,000 in other non-depreciable materials that will be purchased during the first year. The annual project income (net) is projected to be $28,000. At the end of the eight years, the project will be discontinued and the equipment sold for $24,000. Assuming a tax rate of 28% and a MARR of 10%, should the project be undertaken?

Solution

Cash expenses are multiplied by (1 - tax rate)
Incomes are multiplied by (1 - tax rate)
Depreciation is multiplied by tax rate
Recaptured depreciation is multiplied by (1 - tax rate)

Year		
0	First Cost	-108,000
1	Other Costs 25,000(P/F, 10%, 1)(.72)	-16,364
1-8	Depr. [24,000(P/A, 10%, 8) - 3,000(P/G, 10%, 8)](.28)	22,387
1-8	Income 28,000(P/A, 10%, 8)	107,554
8	Recaptured Depr. 24,000(P/F, 10%, 8)(.72)	8,061
	NPW	$13,638

Project should be undertaken.

12-19

The Salsaz-Hot manufacturing company must replace a machine used to crush tomatoes for its salsa. The industrial engineer favors a machine called the Crusher. Information concerning the machine is presented below.

First Cost	$95,000
Annual Operating Costs	6,000
Annual Insurance Cost*	1,750
Annual Productivity Savings	19,000

*Payable at the beginning of each year

Depreciable Salvage Value	$10,000
Actual Salvage Value	14,000
Depreciable Life	6 yrs
Actual Useful Life	10 yrs
Depreciation Method	STL

Property taxes equal to 5% of the first cost are payable at the end of each year.

Relevant financial information for Salsaz-Hot:

Marginal Tax Rate	34%
MARR	10%

Determine net present worth.

Solution

Cash expenses are multiplied by (1 - tax rate)
Incomes are multiplied by (1 - tax rate)
Depreciation is multiplied by tax rate
Capital recovery is not taxable therefore multiplied by 1
Recaptured depreciation is multiplied by (1 - tax rate)

Year		
0	First Cost	-95,000
1-10	Net Savings 13,000(P/A, 10%, 10)(.66)	52,724
1-10	Property Taxes .05(95,000)(P/A, 10%, 10)(.66)	-19,265
0-9	Insurance [1,750 + 1,750(P/A, 10%, 9)](.66)	-7,807
1-6	Depr. 14,167(P/A, 10%, 6)(.34)	20,977
10	Capital Recovery 10,000(P/F, 10%, 10)	3,855
10	Recaptured Depr. 4,000(P/F, 10%, 10)(.66)	1,018
	NPW	-$43,498

12-20

Momma Mia's Pizza must replace its current pizza baking oven. The two best alternatives are presented below.

	Crispy Cruster	Easy Baker
Initial Cost	$24,000	$33,000
Annual Costs	9,000	6,000
Depreciable Salvage Value	6,000	5,000
Actual Salvage Value	6,000	8,000
Depreciable Life	3 yrs	4 yrs
Actual Useful Life	5 yrs	5 yrs
Depreciation Method	SL	SOYD

Assume Momma pays taxes at the 34% rate and has a MARR of 8%. Which oven should be should be chosen?

Solution

CC

Year			
0	First Cost		-24,000
1-5	Annual Costs 9,000(P/A, 8%, 5)(.66)		-23,718
1-3	Depr. 6,000(P/A, 8%, 3)(.34)		5,257
5	Capital Recovery 6,000(P/F, 8%, 5)		4,084
		NPW	-$38,377

EB

Year			
0	First Cost		-33,000
1-5	Annual Costs 6,000(P/A, 8%, 5)(.66)		-15,813
1-4	Depr. [11,200(P/A, 8%, 4) - 2,800(P/G, 8%, 4)](.34)		8,185
5	Capital Recovery 5,000(P/F, 8%, 5)		3,403
5	Recaptured Depreciation 3,000(P/F, 8%, 5)(.66)		1,348
		NPW	-$35,877

Choose the Easy Baker oven.

12-21

A heat exchanger purchased by Hot Spot Manufacturing cost $24,000. The exchanger will save $4,500 in each of the next 10 years. Hot Spot will use SOYD depreciation over a six year depreciable life. The declared salvage value is $3,000. It is expected the exchanger can actually be sold for the declared value. Hot Spot pays taxes at a combined rate of 42% and has a MARR of 8%. Was the purchased justified?

Solution

Year		
0	First Cost	-24,000
1-10	Annual Savings 4,500(P/A, 8%, 10)(.58)	17,513
1-6	Depr. [6,000(P/A, 8%, 6) - 1,000(P/G, 8%, 6)](.42)	7,230
10	Capital Recovery 3,000(P/F, 8%, 10)	1,390
	NPW	$2,133

Yes the purchase was justified.

12-22

Pinion Potato Chip Inc., must purchase new potato peeling equipment for its Union City, Tennessee plant. The plant engineer, Abby Wheeler, has determined that there are three possible set-ups that could be purchased. Relevant data are presented below.

	Naked Peel	Skinner	Peel-O-Matic
First Cost	$45,000	$52,000	$76,000
Annual Costs	6,000	5,000	5,000
Declared Salvage Value	12% of FC	5,500	10,000
Useful Life	6 years	6 years	6 years
Actual Salvage Value	6,500	5,500	12,000

Part A All assets are depreciated using the SL method. Determine which set-up should be chosen if P^2C Inc. has a MARR of 10% and pays taxes at the 34% marginal rate.

Part B Due to economic considerations P^2C Inc. must eliminate from consideration the Peel-O-Matic set-up and because of a change in the tax laws they must use MACRS depreciation. If all other information concerning the other two alternatives remains the same, which should be chosen?

Solution

Part A

NP

Year			
0	First Cost		-45,000
1-6	Annual Costs 6,000(P/A, 10%, 6)(.66)		-17,246
1-6	Depr. 6,600(P/A, 10%, 6)(.34)		9,773
6	Capital Recovery 5,400(P/F, 10%, 6)		3,048
6	Recaptured Depr. 1,100(P/F, 10%, 6)(.66)		410
		NPW	-$49,015

S

Year			
0	First Cost		-52,000
1-6	Annual Costs 5,000(P/A, 10%, 6)(.66)		-14,372
1-6	Depr. 7,750(P/A, 10%, 6)(.34)		11,475
6	Capital Recovery 5,500(P/F, 10%, 6)		3,105
		NPW	-$51,792

POM

Year			
0	First Cost		-76,000
1-6	Annual Costs 5,000(P/A, 10%, 6)(.66)		-52,724
1-6	Depr. 11,000(P/A, 10%, 6)(.34)		16,288
6	Capital Recovery 10,000(P/F, 10%, 10)		5,645
6	Recaptured Depr. 2,000(P/F, 10%, 10)(.66)		745
		NPW	-$67,694

Choose Naked Peel.

Part B

NP

Year			
0	First Cost		-45,000
1-6	Annual Costs 6,000(P/A, 10%, 6)(.66)		-17,246
1	Depr. (.3333)(45,000)(P/F, 10%, 1)(.34)		4,636
2	Depr. (.4445)(45,000)(P/F, 10%, 2)(.34)		5,620
3	Depr. (.1481)(45,000)(P/F, 10%, 3)(.34)		1,702
4	Depr. (.0741)(45,000)(P/F, 10%, 4)(.34)		774
6	Recaptured Depr. 6,500(P/F, 10%, 6)(.66)		2,422
		NPW	-$47,092

S

Year			
0	First Cost		-52,000
1-6	Annual Costs 5,000(P/A, 10%, 6)(.66)		-14,372
1	Depr. (.3333)(52,000)(P/F, 10%, 1)(.34)		5,357
2	Depr. (.4445)(52,000)(P/F, 10%, 2)(.34)		6,494
3	Depr. (.1481)(52,000)(P/F, 10%, 3)(.34)		1,967
4	Depr. (.0741)(52,000)(P/F, 10%, 4)(.34)		895
6	Recaptured Depr. 5,500(P/F, 10%, 6)(.66)		2,049
		NPW	-$49,610

Choose Naked Peel.

Chapter 13

Replacement Analysis

13-1

One of the four ovens at a bakery is being considered for replacement. Its salvage value and maintenance costs are given in the table below for several years. A new oven costs $80,000 and this price includes a complete guarantee of the maintenance costs for the first two years, and it covers a good proportion of the maintenance costs for years 3 and 4. The salvage value and maintenance costs are also summarized in the table.

	Old Oven		New Oven	
Year	Salvage value at end of year	Maintenance costs	Salvage value at end of year	Maintenance costs
0	$20,000	$ --	$80,000	$ --
1	17,000	9,500	75,000	0
2	14,000	9,600	70,000	0
3	11,000	9,700	66,000	1,000
4	7,000	9,800	62,000	3,000

Both the old and new ovens have similar productivities and energy costs. Should the oven be replaced this year, if the MARR equals 10%?

Solution

The old oven ("defender")

Year	S value at end-of-year	EAC Capital Recovery (P-S)x(A/P,10%,n) + Si	Maint. costs	EAC Maintenance = 9,500 + 100(A/G,10%,n)	EAC Total
0	P = 20,000	---	--	--	--
1	17,000	5,000.00	9,500	9,500.00	14,500.00
2	14,000	4,857.20	9,600	9,547.60	14,404.80
3	11,000	4,718.90	9,700	9,593.70	14,312.60*
4	7,000	4,801.50	9,800	9,638.10	14,439.60

*Economic life = 3 years, with EAC = $14,312.60

The new oven ("challenger")

Year	S value at end-of-year	EAC Capital Recovery (P-S)x(A/P,10%,n) + Si	Maint. costs	EAC Maintenance = 9,500 + 100(A/G,10%,n)	EAC Total
0	P = 80,000	---	--	--	--
1	75,000	13,000.00	0	0	13,000.00
2	70,000	12,762.00	0	0	12,762.00
3	66,000	12,229.40	1,000	302.10[a]	12,531.50*
4	62,000	11,879.00	3,000	883.55[b]	12,762.55

[a] 1,000(A/F, 10%, 3) = $302.10
[b] [1,000(F/P, 10%, 1) + 3000](A/F, 10%, 4) = $883.55
*Economic life = 3 years, with EAC = $12,531.50

Since EAC defender > EAC challenger (14,312.6 > 12,531.5) replace oven this year.

13-2
The cash flow diagram below indicates the costs associated with a piece of equipment. The investment cost is $5000 and there is no salvage. During the first 3 years the equipment is under warranty so there are no maintenance costs. Then the estimated maintenance costs. Then the estimated maintenance costs over 15 years follow the pattern shown. To show you can do the calculations required to find the most economic useful life, determine the equivalent annual cost (EAC) for n = 12 if the minimum attractive rate of return (MARR) = 15%. You must use gradient and uniform series factors in your solution.

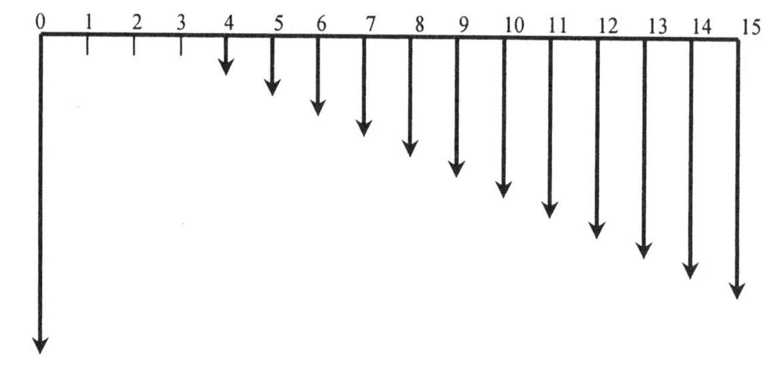

Solution

EAC = 5,000(P/A, 15%, 12) + 150(F/A, 15%, 9)(A/F, 15%, 12)
+ 100(P/G, 15%, 7)(P/F, 15%, 5)(A/P, 15%, 12)
= $1,103

13-3

A hospital is considering buying a new $40,000 diagnostic machine that will have no salvage value after installation, as the cost of removal equals any sale value. Maintenance is estimated at $2000 per year as long as the machine is owned. After ten years the radioactive ion source will have caused sufficient damage to machine components that safe operation is no longer possible and the machine must be scrapped. The most economic life of this machine is:
(Select one)

(a) One year since it will have no salvage after installation.
(b) Ten years because maintenance doesn't increase.
(c) Less than ten years but more information is needed to determine it.

Solution

The correct answer is (b).

13-4

A petroleum company, whose minimum attractive rate of return is 10%, needs to paint the vessels and pipes in its refinery periodically to prevent rust. "Tuff-Coat", a durable paint, can be purchased for $8.05 a gallon while "Quick-Cover", a less durable paint, costs $3.25 a gallon. The labor cost of applying a gallon of paint is $6.00. Both paints are equally easy to apply and will cover the same area per gallon. Quick-Cover is expect to last 5 years. How long must Tuff-Coat promise to last to justify its use?

Solution

This replacement problem requires that we solve for a breakeven point. Let N represent the number of years Tuff-Coat must last. The easiest measure of worth to use in this situation is equivalent annual cash flow (EACF). Although more computationally cumbersome, others could be used and if applied correctly they would result in the same answer.

Find N such that $EACF_{TC} = EACF_{QC}$

$$14.05(A/P, 10\%, N) = 9.25\ (A/P, 10\%, 5)$$

$$(A/P, 10\%, N) = 0.17367$$

Therefore N = 9 years

Tuff-Coat must last at least 9 years. Notice that this solution implicitly assumes that the pipes need to be painted indefinitely (i.e., forever) and that the paint and costs of painting never change (i.e., no inflation or technological improvements affecting the paint or the cost to produce and sell paint, or to apply the paint).

13-5

Ten years ago Hyway Robbery, Inc. installed a conveyor system for $8000. The conveyor system has been fully depreciated to its zero salvage value. The company is considering replacing the conveyor because maintenance costs have been increasing. The estimated end-of-year maintenance costs for the next five years are as follow:

Year	Maintenance
1	$1,000
2	1,250
3	1,500
4	1,750
5	2,000

At any time the cost of removal just equals the value for scrap metal. The replacement the company is considering has an equivalent annual cost (EAC) of $1028 at its most economic life. The company requires a minimum attractive rate of return (MARR) of 10%.

(a) Should the conveyor be replaced now? Show the basis used for the decision.
(b) Now assume the old conveyor could be sold at any time as scrap metal for $500 more than the cost of removal. All other data remain the same. Should the conveyor be replaced?

Solution

a) Since the current value ($0.00) is not changing but maintenance costs are increasing, the most economic life is one year.

Year	Cash Flow
0	0
1	-1,000
S	0

Defender uniform equivalent cost: $EAC_D = \$1,000$

Since $EAC_D < EAC_C$, keep the old conveyor for now.

b)

Year	Cash Flow
0	-500
1	-1,000
S	+500

$EAC_D = 1,000 + 500(A/P, 10\%, 1) - 500(A/F, 10\%, 1) = \$1,050.$

Since $EAC_D > EAC_C$, replace the old conveyor.

13-6

Ten years ago, the Cool Chemical Company installed a heat exchanger in its plant for $10,000. The company is considering replacing the heat exchanger because maintenance cost have been increasing. The estimated maintenance costs for the next 5 years are as follow:

Year	Maintenance
1	$1,000
2	1,200
3	1,400
4	1,600
5	1,800

Whenever the heat exchanger is replaced, the cost of removal will be $1,500 more than the heat exchanger is worth as scrap metal. The replacement the company is considering has an equivalent annual cost (EAC) = $900 at its most economic life. Should the heat exchanger be replaced now if the company's minimum attractive rate of return (MARR) is 20%?

Solution

Since the current value ($-1,500) is not changing but maintenance costs are increasing, the most economic life is one year.

Year	Cash Flow	
0	+1,500	(Foregone salvage)
1	-1,000	(Maintenance)
S	-1,500	(Negative Salvage)

Equivalent annual cost of the defender:
$EAC_D = 1{,}000 + 1{,}500(A/F, 20\%, 1) - 1{,}500(A/P, 20\%, 1) = \$700$

Since $EAC_D < EAC_C$, keep the old heat exchanger for now.

13-7

An engineer is trying to determine the economic life of new metal press. The press costs $10,000 initially. First year maintenance cost is $1,000. Maintenance cost is forecast to increase $1,000 per year for each year after the first. Fill in the table below and determine the economic life of the press. Consider only maintenance and capital recovery in your analysis. Interest is 5%.

Year	Maintenance Cost	EAC of Capital Recovery	EAC of Maintenance	Total EAC
1	$1,000			
2	2,000			
3	3,000			
4	4,000			
5	5,000			
6	6,000			
7	7,000			
8	8,000			

*EAC = Equivalent Annual Cost

Solution

Year	Maintenance Cost	EAC of Capital Recovery	EAC of Maintenance	Total EAC
1	$1,000	$11,500	$1,000	$12,500
2	2,000	6,151	1,465	7,616
3	3,000	4,380	1,907	6,287
4	4,000	3,503	2,326	5,829
5	5,000	2,983	2,723	5,706
6	6,000	2,642	3,097	5,739
7	7,000	2,404	3,450	5,854
8	8,000	2,229	3,781	6,010

Economic Life = 5 yrs (EAC = minimum)

EAC of Capital Recovery = $10,000 (A/P, 15%, n)
EAC of Maintenance = $1,000 + 1,000 (A/G, 15%, n)

13-8

A manufacturer is contemplating the purchase of an additional forklift truck to improve material handling in the plant. He is considering two popular models, the Convair T6 and the FMC 340. The relevant financial data are shown below. His minimum attractive rate of return (MARR) is 12%.

Model	First Cost	Life	Salvage Value	Annual Operating Cost
Convair T6	$20,000	6 years	$2,000	$8,000
FMC 340	29,000	7 years	4,000	4,000

(a) Which model is more economical?
(b) List two important assumptions that are implicit in your computations in (a).

Solution

(a) Compute the EAW for each model.

Convair: EAW = -20,000(A/P, 12%, 6) + 2,000(A/F, 12%, 6) - 8,000
= -$12,617.60

FMC: EAW = -29,000(A/P, 12%, 7) + 4,000(A/F, 12%, 7) - 4,000
= -$9,957.50

The FMC is more economical.

b) That either truck can be
(1) repeated identically into the indefinite and
(2) the service to be provided (material handling) is required forever.

13-9

A graduate of an engineering economy course has complied the following set of estimated costs and salvage values for a proposed machine with a first cost of $15,000; however, he has forgotten how to find the most economic life. Your task is to show him how to do this by calculating the equivalent annual cost (EAC) for n = 8, if the minimum attractive rate of return (MARR) is 15%. You must show how to use gradients and uniform series factors in your solution.

Life (n) Years	Estimated End-of-Year Maintenance	Estimated Salvage if sold in Year n
1	$ 0	$10,000
2	$ 0	9,000
3	300	8,000
4	300	7,000
5	800	6,000
6	1,300	5,000
7	1,800	4,000
8	2,300	3,000
9	2,800	2,000
10	3,300	1,000

Remember: Calculate only <u>one</u> EAC (for n = 8). You are not expected to actually find the most economical life.

Solution

First Cost	EAC = 15,000(A/P, 15%, 8) = \$3,344
Salvage Value	EAC = -3,000(A/F, 15%, 8) = -\$219
Maintenance	EAC = 300(F/A, 15%, 6)(A/F, 15%, 8) + 500(P/G, 15%, 5)(P/F, 15%, 3)(A/P, 15%,8) = \$615
Total	EAC_8 = \$3,740

(A complete analysis would show that the most economic life is 7 years, with EAC = \$3,727)

13-10

An existing machine has operating costs of \$300 per year and a salvage value of \$100 (for all years). A new replacement machine would cost \$1,000 to purchase, and its operating cost over the next period of t years (not per year) is $M = 200t + 10t^2$. Assume i = zero percent.

(a) What is the most economic life t* for the new machine?
(b) Should the old machine be replaced with the new one?

Solution

(a) Cost per year = $AC = \frac{1,000}{t} + \frac{M}{t} = \frac{1,000}{t} + 200 + 10t$

$$\frac{dAC}{t} = \frac{1,000}{t^2} + 10 = 0 \Rightarrow t^* = +10 \text{ years}$$

b) $AC^* = AC(10) = \frac{1,000}{10} + 200 + 10(10)$

$= 400 \geq$ Annual cost of old machine for any number of years

∴ No, keep old one.

13-11

The computer system used for production and administration control at a large cannery is being considered for replacement. Of the available replacement system, "Challenger I" has been considered the best. However, it is anticipated that after one year, the "Challenger II" model will become available, with significant technological modifications. The salvage value projections for these three systems are summarized below. Assuming that their performance would otherwise be comparable, should we replace the existing system either this year or next year? Assume the MARR equals 12%, and a useful life of 5 years on all alternatives.

Year	Existing Computer	Salvage value at end of the year Challenger I	Challenger II
0	$20,000	$25,000	$24,000
1	16,000	22,000	23,000
2	13,000	21,000	23,000
3	11,000	20,000	22,000
4	8,000	16,000	16,000
5	3,000	10,000	10,000

Solution

1. Existing Computer (defender)

Year	Salvage value at end-of-year	EAC Capital Recovery = (P - S)(A/P, 12%, n) + Si	
0	P = 20K	--	
1	16K	6,400.00	*Economic life = 4 yrs EAC = $4,910.40
2	13K	5,701.90	
3	11K	5,066.70	
4	8K	4,910.40*	
5	3K	5,075.80	

2. "Challenger I"

Year	Salvage value at end-of-year	EAC Capital Recovery = (P - S)(A/P, 12%, n) + Si	
0	P = 25K	--	
1	22K	6,000.00	*Economic life = 3 yrs EAC = $4,481.50
2	21K	4,886.80	
3	20K	4,481.50*	
4	16K	4,882.80	
5	10K	5,361.00	

3. "Challenger II"

Year[1]	Salvage value at end-of-year	EAC Capital Recovery = (P - S)(A/P, 12%, n) + Si	
0	P = 24K	--	
1	23K	3,880.00	*Economic life = 2 yrs EAC = $4,910.40
2	23K	3,351.70*	
3	22K	3,472.60	
4	16K	4,553.60	
5	10K	5,083.60	

[1]Year numbers do not refer to same time scale as for Challenger 1

Note $EAC_{Challenger\ II} < EAC_{Challenger\ I} < EAC_{Defender}$, but should we replace it now or should we wait one year for the Challenger II?

Alternative A: Don't wait

$EAC_A = EAC_{Challenger\ I} = \$4{,}481.50$

Alternative B: Wait one year for replacement.

$EAC_B = [6{,}400(P/A, 12\%, 1) + (3{,}351{,}7(P/A, 12\%, 2)(P/F, 12\%, 1)](A/P,12\%,3)$

cost of keeping one more year — cost of challenger II at its best (2 yrs economic life)

$= \$4{,}484.49$

Since $EAC_A \approx EAC_B$ we should preferably wait one year, although strictly speaking we can choose either option.

13-12

A truck salesman is quoted as follows:

"Even though our list price has gone up to \$42,000, I'll sell you a new truck for the old price of only \$40,000, an immediate savings of \$2,000, and give you a trade-in allowance of \$21,000, so your cost is only (\$40,000 - 21,000) = \$19,000. The book value of your old truck is \$12,000, so you're making an additional (\$21,000 - 12,000) = \$9,000 on the deal." The salesman adds, "Actually I am giving you more trade-in for your old truck than the current market value of \$19,500, so you are saving an extra (\$21,000 - 19,500) = \$1,500."

(a) In a proper replacement analysis, what is the first cost of the defender?
(b) In a proper replacement analysis, what is the first cost of the challenger?

Solution

(a) \$19,500

The defender 1st cost is always the current market value, not trade-in or book value.

(b) \$38,500

With an inflated trade-in of \$1,500 (21,000 - 19,500), the new truck can be purchased for \$40,000. Therefore, the appropriate value for a replacement analysis is: \$40,000 - \$1,500 = \$38,500.

Chapter 14

Inflation and Price Change

14-1

A European investor lives near to one of his country's borders. In Country A (Where he lives), an 8% interest rate is offered in banks, and the inflation rate is 3%. Country B, on the other hand, has an inflation rate of 23%, and their banks are offering 26% interest on deposits.

(a) What is the real or effective interest rate that this person gets when investing in his Country?

(b) This investor believes that the currency of Country B will not change in its value relative to the value of the currency of Country A during this year. In which country would he get a larger effective interest rate?

(c) Suppose now that he invests in a bank in Country B, and that his prediction was wrong. The currency of Country B was devaluated 20% with respect to the exchange value of Country A's currency. What is the effective interest rate that he obtained?

Solution

(a) $i' = ?$ if $i = 80\%$, $f = 3\%$

$$i = i' + f + i'F$$
$$.08 = i' + .03 + i'(.03)$$
$$i' = 0.0485$$
$$= 4.85\%$$

b) if investment in Country A: $i'_A = 0.0485$
if investment in Country B: $i_B = 26\%$, $f_A = 3\%$
(note that he lives in Country A. Inflation of Country B does not affect him directly)

$$i'_B = \frac{i_B - f_A}{1 + f_A} = \frac{0.26 - 0.03}{1 + 0.03} = 0.2233 = 22.33\%$$

He can get a larger effective interest rate in Country B.

(c) Let X = amount originally invested in B (measured in currency A).

The amount collected at end of 1 year (measured in currency A) =

$$\underbrace{(1.0 - 0.2)}_{\text{Due to the devaluation}} \quad \underbrace{(1.26)}_{\text{Due to initial deposit (+) interest}} = 1.008X$$

the interest is then $i = \frac{1.008X - X}{X} = 0.008$

but during the year the inflation in Country A (where he lives) was 3%, therefore

$i = 0.008$
$f = 0.03$
$i' = ?$

$$i' = \frac{0.008 - 0.03}{1 + 0.03} = -0.02136$$

He actually lost money (negative effective interest rate of - 2.136%).

14-2

The first sewage treatment plant for Athens, Georgia cost about $2 million in 1964. The utilized capacity of the plant was 5 million gallons/day (mgd). Using the commonly accepted value of 135 gallons/person/day of sewage flow, find the cost per person for the plant. Adjust the cost to 1984 dollars with inflation at 6%. What is the annual capital expense per person if the useful life is 30 years and the value of money is 100%.

Solution

Population equivalents = 5 mgd/135 = 37,037

$$\text{Cost per capita} = \frac{\$2{,}000{,}000}{37{,}037} = \$54$$

$1984_{\$}$, F = 54(F/P, 6%, 20) = $173.18

Annual Cost, A = 173.18(A/P, 10%, 30) = $18.37

14-3

How much life insurance should a person buy if he wants to leave enough money to his family, so they get $25,000 per year in interest, of consent Year 0 value dollars? The interest rate expected from banks is 11%, while the inflation rate is expected to be 4% per year.

Solution

The actual (effective) rate that the family will be getting is

$$i' = \frac{i-f}{1+f} = \frac{0.11-0.04}{1.04} = 0.0673 = 6.73\%$$

To calculate P, $n = \infty$ (capitalized cost)

$$P = \frac{A}{i'} = \frac{25{,}000}{0.0673} = \$371{,}471$$

Therefore, he needs to buy about 370,000 of life insurance.

14-4

Property, in the form of unimproved land, is purchased at a cost of $8000 and is held for six years when it is sold for $32,600. An average of $220 each year is paid in property tax and may be treated at an interest of 12%. The long-term capital gain tax is 15% of the long-term capital gain tax is 15% of the long-term capital gain. Inflation during the period is treated as 7% per year. What is the annual rate of return for this investment?

Solution

Long term gains = 32,600 – 8,000 = 24,600
Tax on long-term gain = .15 x 24,600 = 3,690
Property tax = 220(F/A, 12%, 6) = 1,785.30

Adjusted FW = 32,600 - 3,690 - 1785.30 = 27,624.70
also $FW = 8{,}000(1 + i_{eq})^6$

$$\therefore \left(1+i_{eq}\right) = \left(\frac{27{,}124.70}{8{,}000}\right)^{\frac{1}{6}} = 1.2257$$

$$\left(1+i_{eq}\right) = \left(1+i\right)\left(1+i_f\right)$$

$$1+i = \frac{1.2257}{1.07} = 1.1455 \text{ or } 14.6\% \text{ rate of retrun}$$

14-5

The auto of your dreams costs $20,000 today. You have found a way to earn 15% tax free on an "auto purchase account". If you expect the cost of your dream auto to increase by 10% per year, how much would you need to deposit in the "auto purchase account" to provide for the purchase of the auto 5 years from now?

Solution

Cost of Auto 5 years hence (F) = P(1 + inflation rate)^n = 20,000 (1 + 0.10)^5 = 32,210

Amount to deposit now to have $32,210 five years hence

P = F(P/F, i%, n) = 32,210(P/F, 15%, 5) = $16,014.81

14-6

On January 1, 1975 the National Price Index was 208.5, and on January 1, 1985 it was 516.71. What was the inflation rate, compounded annually, over that 10-year period? If that rate continues to hold for the next 10 years, what National Price Index can be expected on January 1,1995?

Solution

$$NPW = 0 - 208.5 + 516.71(P/F, i_f, 10)$$

$$(P/F, i_f, 10) = \frac{208.5}{516.71} = 0.4035$$

Trial & Error Solution:

Try i = 9%: $-208.5 + 516.71(0.4224) = +9.76$

Try i = 10%: $-208.5 + 516.71(0.3855) = -9.31$ $\qquad 9\% < i_f < 10\%$

By interpolation $i_f = 9.51\%$

National Price $\text{Index}_{1995} = 516.71(1 + 0.0951)^{10} = 1{,}281.69$

14-7

A department store offers two options to buy a new color TV which has a price of $440.00. A customer can either pay cash and receive immediately a discount of $49.00 or he can pay for the TV on the installment plan. The installment plan has a nominal rate of 12% compounded bi-yearly and would require an initial down payment of $44.00 followed by four equal payments (principle and interest) every six months for two years.

If for the typical customer the real minimum attractive rate of return is 5%, what is the maximum effective annual inflation rate for the next two years that would make paying cash preferred to paying installments? All figures above are quoted in time zero dollars.

Solution

The monthly payments in nominal dollars if the installment plan was selected would be

$$(-440 + 44)(A/P, 6\%, 4) = -\$114.28$$

The breakeven inflation rate is that such that $NPV_{BUY} = NPV_{INSTALL}$ or $NPV_{BUY - INSTALL} = 0$

$$NPV_{B-I} = ((-440 + 49) + 44) + 114.28(P/A, i_{1/2}, 4) = 0$$

$(P/A, i_{1/2}, 4) = 3.0364$ therefore the nominal effective semi-annual MARR would have to be $i_{1/2} = .115$. The nominal effective annual rate would be $i = (1.115)^2 - 1 = 0.2432$

The effective annual inflation rate can now be computed from the formula

$$(1.2432) = (1.05)(1 + f)$$
$$f = .1840$$

14-8

An automobile that cost $16,500 in 1998 has an equivalent model four years later in 2002 that cost $19,250. If inflation is considered the cause of the increase, what was the average annual rate of inflation?

Solution

$$F = P(1 + i_f)$$
$$19{,}250 = 16{,}500(1 + i_f)$$
$$\frac{19{,}250}{16{,}500} = (1 + i_f)$$
$$1 + i_f = (1.167)^{1/4}$$
$$= 1.039$$
$$i_f = 3.9\%$$

14-9

A machine has a first cost of $100,000 (in today's dollar) and a salvage value of $20,000 (in then current dollars) at the end of its ten year life. Each year it will eliminate one full-time worker. A worker costs $30,000 (today's dollars) in salary and benefits. Labor costs are expected to escalate at 10% per year. Operating and maintenance costs will be $10,000 per year (today's dollars) and will escalate at 7% per year.

Construct a table showing before-tax cash flows in current dollars, and in today's dollars. The inflation rate is 7%.

Solution

End of Year	Current Dollars				Today's Dollar's
	Savings	O & M	Capital	Total	
0			-100,000	-100,000	-100,000
1	33,000	-10,700		22,300	20,841
2	36,300	-11,449		24,851	21,706
3	39,930	-12,250		27,680	22,595
4	43,923	-13,108		30,815	23,509
5	48,315	-14,026		34,290	24,448
6	53,147	-15,007		38,140	25,414
7	58,462	-16,058		42,404	26,407
8	64,308	-17,182		47,126	27,428
9	70,738	-18,385		52,354	28,477
10	77,812	-19,672	20,000	78,141	39,723

14-10

A project has been analyzed assuming 6% inflation and is found to have a monetary internal rate of return (IRR) of 22%. What is the real IRR for the project?

Solution

Real IRR = (1.22)/(1.06) - 1 = 0.1509 or 15.09%

14-11

A company requires a real MARR of 12%. What monetary MARR should they use if inflation is expected to be 7%?

Solution

Monetary MARR = (1.12)(1.07) - 1 = 0.1984 or 19.84%

14-12

The real interest rate is 4%. The inflation rate is 8%. What is the apparent interest rate?

Solution

$$i = i' + f + i'f$$
$$= 0.04 + 0.08 + 0.04(0.08) = 12.32\%$$

14-13

A solar energy book gives values for a solar system as follows: Initial cost, \$6,500; Initial fuel savings, \$500/year; Expected life, 15 years; Value of Money, 10%; Inflation, 12%; and Incremental income tax rate, 25%. If we define the payback condition as the time required for the present worth of the accumulated benefit to equal the accumulated present worth of the system cost, what is the time required to reach the payback condition? Since the income tax benefit is related to the annual interest expense, treat it as a reduction of the annual cost.

Solution

Annualizing P: A = 6,500(A/P, 10%, 15)
= 854.75

$1 + i_C = (1.10)(1 + 0.25 \times 0.10) = 1.1275$

PW of costs = 854.75(P/A, 12.75%, 15) = 5,595.82

$$1 + i_{eq} = \frac{1 + i_f}{1 + i} = \frac{1.12}{1.10} = 1.018$$

The solution strategy is to find the time for the PW of benefits to equal PW of cost. When the combined effect of the two rates on a distributed A amount are opposed then the net effect retains the direction of the longer rate. The inflation rate is greater than the time value of money, which is abnormal. To solve this problem, find the PW of benefit, and to do that we must get FW of the equivalent rate, i_{eq}.

Try 10 years: FW = 500(F/A, 1.8%, 10) = 500 (10.850) = \$5,425.06
Try 11 years: FW = 500(F/A, 1.8%, 11) = 500 (12.045) = \$6,022.72

10 years < Payback < 11 years By interpolation payback = 10.3 years

14-14

Compute the internal rate of return based on constant (Year 0) dollars for the following after-tax cash flow given in current or actual dollars. Inflation is assumed to be 7% per year. (Round to the nearest dollar).

Year	After Tax Cash Flow In Actual Dollars
1998 (Yr 0)	-\$10,000
1999	3,745
2000	4,007
2001	4,288
2002	4,588

Solution

Year	After Tax Cash Flow In Constant Dollars
1998 (0)	-\$10,000
1999 (1)	$3,745(1.07)^{-1} = 3,500$
2000 (2)	$4,007(1.02)^{-2} = 3,500$
2001 (3)	$4,288(1.07)^{-3} = 3,500$
2002 (4)	$4,588(1.07)^{-4} = 3,500$

NPW = 0 at IRR

$$0 = -10,000 + 3,500(P/A, i\%, 4)$$
$$(P/A, i\%, 4) = 10,000/3,500 = 2.857$$

Searching tables where n = 4, i = IRR = 15%

14-15

The capital cost of a wastewater treatment plant for a small town of 6,000 people was estimated to be about \$85/person in 1969. If a modest estimate of the rate of inflation is 5.5% for the period to 1984, what is the per capita capital cost of a treatment plant now?

Solution

$$F = P(1+ i_f)^n = 85(1+.055)^{15} = 85(2.232) = \$189.76$$

14-16

A lot purchased for \$4,500 is held for five years and sold for \$13,500. the average annual property tax is \$45 and may be accounted for at an interest rate of 12%. The income tax on the long term capital gain is at the rate of 15% of the gain. What is the rate of return on the investment if the allowance for inflation is treated at an average annual rate of 7%?

Solution

Long term gain = 13,500 - 4,500 = 9,000
Tax on long term gain = (.15)(9,000) = 1,350
Property tax = 45(F/A, 12%, 5) = 285.89

Adjusted FW = 13,500 - 1,350 - 285.89 = 11,864.12
also FW = $4{,}500(1 + i_{eq})^5$

$$\therefore \left(1+i_{eq}\right) = \left(\frac{11{,}864.12}{4{,}500}\right)^{\frac{1}{5}} = 1.214$$

$$\left(1+i_{eq}\right) = (1+i)\left(1+i_f\right)$$

$$1+i = \frac{1.214}{1.075} = 1.129 \text{ or } 12.9\% \text{ rate of retrun}$$

14-17

Undeveloped property near the planned site of an interest highway is established to be worth $48,000 in six years when the construction of the highway is completed. Consider a 15% capital gains tax on the gain, an annual property tax of 0.85% of the purchase price, an annual inflation rate of 7%, and an expected return of 15% on the investment. What is the indicated maximum purchase price now?

Solution

Let X = purchase cost

$1 + i_{eq} = (1.15)(1.07) = 1.231$

Annual property tax = .0085X
FW of property tax = .0085X(F/A, 23.1%, 6) = .0909X

Adjusted return = 48,000 - .15(48,000 - X) - .0909X

Also = $X(1.231)^6 = 3.48X$

Therefore 40,800 + .15 X - .0909X = 3.48X
X = $11,927 purchase cost

14-18
Jack purchases a lot for $40,000 cash and plans to sell it after 5 years. What should he sell it for if he wants a 20% before-tax rate of return, after taking the 5% annual inflation rate into account?

Solution

F = 40,000(F/P, 20%, 5)(F/P, 5%,5)
= $126,988

14-19
Minor Oil Co. owns several gas wells and is negotiating a 10-year contract to sell the gas from these wells to Major Oil Co. They are negotiating on the price of the gas the first year, $ per thousand cubic feet (KCF), and on the escalation clause, the percentage rate of increase in the price every year thereafter. Minor experts the wells to produce 33,000 KCF the first year and to decline at the rate of 18% every year thereafter. Minor has agreed to spend $500,000 now to lay pipelines from each well to Major's nearby refinery. What should the minimum price be the first year and what should the escalation rate be if Minor wants their revenue each year to remain constant (uniform) over the life of the contract. Assume an end-of-year convention and a minimum attractive rate of return (MARR) of 15%.

Solution

Required income to earn the 15% MARR on $500,000:

EAB = 500,000(A/P, 15%, 10) = = $99,650.

First year price = $99,650/33,000KCF = $3.02/KCF

Annual production declines to (1 - 0.18) of initial rate each year.

Let f = required annual escalation rate

Then (1 - 0.18)(1+f) = 1 to keep the revenue constant

$$f = \frac{1}{(1-0.18)} - 1$$

= 0.2195/year

14-20
A solar system costs $6500 initially and qualifies for a federal tax credit (40% of cost, not to exceed $4,000). The cost of money is 10%, and inflation is expected to be 7% during the life of the system. The expected life of the system is 15 years with zero salvage value. The homeowner is in the 40% income tax bracket. The initial fuel saving is estimated at $500 for the first year and will increase in response to inflation. The annual maintenance cost of the system is established at 5% of the annualized cost of the system. What is the time required for the payback condition to be reached for this investment?

Solution

Adjust initial cost by tax credit: $P = .60(6{,}500) = 3{,}900$

Annualized cost: $A = 3{,}900(A/P, 10\%, 15) = 512.85$

$1 + i_c = \frac{1.10(1+.40(.10))}{(1.05)} = 1.0895$ $\quad$ $1 + i_m = 1.05$ represents maintenance charge as a rate

PW of costs = 512.85(P/A, 8.95%, 15) = 512.85(8.086) = 4,146.67

$1 + i_{eq} = (1 + i)/(1 + i_f) = 1.10/1.07 = 1.028$

Try 9 years: PW = 500 (P/A, 2.8%, n) = 500 (7.868) = \$3,934.18
Try 10 years: PW = 500 (P/A, 2.8%, n) = 500 (8.618) = \$4,308.97

9 years < Payback < 10 years $\quad$ By interpolation payback = 9.6 years

14-21
Acme Company is considering the purchase of a new machine for \$30,000 with an expected life of 20 years when it is established the salvage value will be zero. An investment tax credit of 7% will be allowed. The incremental tax rate (tax bracket) for the company is 36%, inflation is established to be 7%, and the value of money is 12%. If the annual benefit is established to be \$2,500 per year over current production and maintenance costs, what will be the time required for the payback condition to be reached (that is, the point where the present worth of benefits up to payback equals the present worth of costs for the life expectancy of the equipment)?

Solution

Tax credit reduces initial cost $P = 30{,}000 \times .93 = 27{,}900$

Annualize P: $A = 27{,}900(A/P, 12\%, 20) = 3735.81$

i_c combines interest i with income tax credit rate
$1 + i_c = (1 + i)(1 + .36 \times .12) = 1.1684$

PW of cost = 3,735.81(P/A, 16.84%, 20) = 3,735.81(5.674) = 21.197

$1 + i_{eq} = (1 + i)/(1 + i_f) = 1.12/1.07 = 1.0467$

Try 11 years: PW = 2,500 (P/A, 4.67%, n) = \$21,130
Try 12 years: PW = 2,500 (P/A, 4.67%, n) = \$22,575

11 years < Payback < 12 years $\quad$ By interpolation payback = 11.05 years

14-22

The net cost of a solar system for a home is \$8,000 and it is expected to last 20 years. If the value of money is 10%, inflation is expected to be 8%, and the initial annual fuel saving is \$750, what is the time for the payback condition to be reached for the system? Assume the homeowner is in the 30% income tax bracket.

Solution

Annualize P: A = 8,000 (A/P, 10%, 20) = 940

$1 + i_c = (1.10)(1 + .10 \times .30) = 1.133$

PW of Cost = 940(P/A, 13.3%, 20) = 940(6.900) = 6,486

$1 + i_{eq} = (1 + i)/(1 + i_f) = 1.10/1.08 = 1.0185$

Try 9 years: PW = 940(P/A, 1.85%, n) = \$6,171
Try 10 years: PW = 940(P/A, 1.85%, n) = \$6,790

9 years < Payback < 10 years By interpolation payback = 9.5 years

14-23

A undeveloped percent of land in Clarke County, Georgia was purchased in 1980 for \$4,850. The property tax was \$8 for the first year and is assumed to have increased by \$ 2 per year. The capital gain tax is 13.6% of long term capital gain. Inflation for the period is treated at an 8% annual rate. A 16% rate of return on the investment is desired. What is the indicated sale price in 1985?

Solution

$1 + i_{eq} = (1.16)(1.08) = 1.2528$

FW of property tax = [8 + 2(A/G, 25.28%, 5)] [F/A, 25.28%, 5]
= [8+ 2(3.12)] [8.252]
= 91.74

Let X = selling price

Long term capital gains tax = 0.136(X - 4,850) = .136X - 659.60

Adjusted return = X - [.136X - 659.60 + 91.74] = .864X + 567.86

Also = 4,850(1.2528) = 14,967.54

.864X + 567.86 = 14,967.54
.864X = 14,399.68
X = \$16,666.31 selling price

14-24

A company has designed a VLSI circuit and a production system to manufacture it. It is believe that it can sell 100,00 circuits per year if the price in then-current dollars is cut 20% per year (for example, if the unit price in the first year is \$100, then the price in years 2 through 5 would be \$80, \$64, \$51.20, and \$40.96).

The required revenue for the five years is \$2,500.00 per year in today's dollars. The real and monetary costs of capital are 8.8% and 16.416% respectively. What should the then-current dollar selling price be in each of the years 1 through 5?

Solution

Let R be the required revenue in year 1, then the required revenue in years 2 through 5 is .8R, .64R, .512R, and .0496R. Since these are in then-current \$,

$$(2{,}500{,}000)(P/A,\ 8.8\%,\ 5) = R(1.16416)^{-1} + 0.8R(1.16416)^{-2} + .064R(1.16416)^{-3}$$
$$+\ 0.512R(1.16416)^{-4} + 0.4096R(1.16416)^{-5}$$
$$9{,}774{,}800 = 2.32523R$$

R = 4,203,804 or a unit price of
\$42.04 in year 1
\$33.63 in year 2
\$26.90 in year 3
\$21.52 in year 4
\$17.22 in year 5

14-25

Steve Luckee has just been informed that he has won \$1 million in a local state lottery; unfortunately the Internal Revenue Service has also been so informed. Due to tax considerations, Steve has a choice in the manner in which he will receive his winnings. He can either receive the entire amount today, in which case his combined federal an, state income tax rate will soar to 74%, or he can receive the money in uniform annual payments of \$50,000 over the next 20 years, beginning one year from today, in which case his annual payments will be taxed at a lower rate. Steve expects inflations to be a problem for the next 20 years which will obviously reduce the value of his winnings if he takes the annual payment option. Steve uses a real dollar MARR of 5% in his decisions.

(a) If Steve estimates that the inflation rate f would be 6% and his combined federal and state tax rate **d** would be 40% for the next 20 years, what action should he take?
(b) What would be the present worth of the cost of an error in Steve's estimates in (a) if the true values eventually prove to be **f** = 8% and **d** = 20%?
(c) If Steve is certain that 6% f 8% and 20% d 40% for the next 20 years, what is the maximum cost of errors in estimating **f** and **d**?

Solution

(a) The value of the entire amount is:
(1 - .74)($1M) = R$260,000.
The real NPV of the payments would be:
NPV = (1 - .4)($1M/20)(P/A, (1.05)(1.06) - 1,20) = R$234,288
Take the entire amount today, it is worth more.

(b) The real NPV of the payments would be:
NPV = (1 - .2)(50K)(P/A, (1.05)(1.08) - 1, 20) = R$274,370
which is more than the entire amount today.
Steve's errors in estimation would cost him
274, 370 - 260,000 = R$14,370.

(c) The maximum cost of an error will occur one of two ways, either Steve's estimates indicates he should take the entire amount today when he shouldn't or that he take the payments when he shouldn't. If he takes the entire amount when shouldn't, then for **f** = .06 and **d** = .2, and the payments are worth
NPV = .8(50K)(P/A, (1.05) (1.06) - 1, 20) = R$312,384,
and the cost is $52,384. If he takes the payments when he shouldn't, then if **f** = .08 and **d** = .4, and the payments are worth .6(50K)(P/A, (1.05)(1.08) - 1, 20) = R$205,777, and the cost is $54,223. The maximum cost is $54,223.

14-26

An investor is considering the purchase of a bond. The bond has a face value of $1000, a coupon rate (interest rate) of i = 6% compounded annually, pays interest once a year, and matures in 8 years. This investor's real MARR is 25%.

(a) If the investor expects an inflation rate of 4% per year for the next 8 years, how much should he be willing to pay for the bond?

(b) If, however, the investor expects a deflation rate of 18% per year for the next 8 years, how much should he be willing to pay for the bond?

Solution

(a) The net cash flows for bond in nominal dollars are

t	Net cash flows
0	N$F.C.
1-8	+60 = .06(1,000)
8	+ 1,000

If the investor wants to earn a real 25%, but expects inflation at 4%, his nominal MARR must be (1.25)(1.04) - 1 = .30, and therefore

NPV = 0 at IRR

0 = -F.C. + 60(P/A, .30, 8) + 1,000(P/F, .30, 8)
F.C. = \$250.50

He should be willing to pay no more than \$177.

(b) If the deflation rate (negative inflation) is 18% and he still expects to earn a real 25% then the real MARR is (1.25)(0.82) - 1 = .025,

0 = -F.C. + 60(P/A, .025, 8) + 1,000(P/F, .025, 8)
F.C. = \$1,251

14-27

An investment in undeveloped land of \$9,000 was held for four years and sold for \$21,250. During this time property tax was paid that was, on the average, 0.4% of the purchase price. Inflation in this time period average 7% and the income tax was 15.2% of the long term capital gain. What rate of return was obtained on the investment?

Solution

Long term capital gains tax = 0.152(21,250 - 9,000) = \$1,862

Property. Tax = .004 x 9,000 = \$36/year

FW of property Tax = 36(F/A, i_{eq}, 4)

$$1 + i_{eq} = \left(\frac{21{,}250-1{,}862}{9000}\right)^{\frac{1}{4}} = 1.2115 \qquad (1^{st} \text{ estimate})$$

FW of property Tax = 36(F/A, 21.15%, 4) = 36(5.47) = \$197

$$1 + i_{eq} = \left(\frac{21{,}250-1{,}862-197}{9000}\right)^{\frac{1}{4}} = 1.2084 \qquad (2^{nd} \text{ estimate})$$

$$\text{Rate of Return} = \left(\frac{1.2084}{1.07}\right) - 1 = 12.9\%$$

Check:

9,000(1.284) =	\$19,191
Property Tax =	197
L.T. Capital Gain Tax =	1,862
	\$21,250

Chapter 15

Selection of a Minimum Attractive Rate of Return

15-1

A small surveying company identifies its available independent alternatives as follows:

Alternative	Initial Cost	Rate of Return
A. Repair existing equipment	$1000	30%
B. Buy EDM instrument	2500	9%
C. Buy a new printer	3000	11%
D. Buy equipment for an additional crew	3000	15%
E. Buy programmable calculator	500	25%

The owner of the company has $5000 of savings currently invested in securities yielding 8% that could be used for the company.

a) Assuming his funds are limited to his savings, what is the apparent cut-off rate of return?
b) If he can borrow money at 10%, how much should be borrowed?

Solution

a)

Alt	Investment	Cumulative Investment	IRR	
A	1,000	1,000	30%	
E	500	1,500	25%	
D	3,000	4,500	15%	← Cut-off rate of return
C	3,000	7,500	11%	= 11% to 15%
B	2,500	10,000	9%	

b) Do all projects with a rate of return > 10%. Thus Alternatives A, E, D, & C with a total initial cost of $7,500 would be selected. Since only $5,000 is available, $2,500 would need to be borrowed.

15-2

The capital structure of a firm is given below.

Source of Capital	Percent of Capitalization	Interest Rate
Loans	35	17%
Bonds	30	8%
Common Stock	35	12%

The combined state and federal income tax rate for the firm is 42%. What is the after-tax and before-tax cost of capital to the firm?

Solution

Before Tax Cost of Capital

0.35 x 17% + 0.30 x 8% + 0.35 x 12% = 12.55%

After Tax Cost of Capital

0.35 x 17% (1-0.42) + 0.30 x 8% (1-0.42) + 0.35 x 12% = 9.04%

15-3

A small construction company identifies the following alternatives that are independent, except where noted.

Alternative	Initial Cost	Incremental Rate of Return	On Investment Over
1. Repair bulldozer	$5,000	30.0%	0
2. Replace Backhoe			
With Model A	20,000	15.0%	0
With Model B	25,000	10.5%	2A
3. Buy new dump truck			
Model X	20,000	20.0%	0
Model Y	30,000	14.0%	3X
4. Buy computer			
Model K	5,000	12.0%	0
Model L	10,000	9.5%	4K

(a) Assuming the company has $55,000 available for investment and it is not able to borrow money, what alternatives should be chosen, and what is opportunity coast of capital?

(b) If the company can also borrow money at 10%, how much should be borrowed, and which alternatives should be selected?

Solution

Rank the alternatives by ΔROR

Project	Incremental Investment	Cumulative Investment	ΔIRR
1	5,000	5,000	30.0%
3X	20,000	25,000	20.0%
2A	20,000	45,000	15.0%
3Y - 3X	10,000	55,000	14.0%
4K	5,000	60,000	12.0%
2B - 2A	5,000	65,000	10.5%
4L - 4K	5,000	70,000	9.5%

a) With $55,000 available choose Projects:

1 Repair bulldozer
2A Backhoe model A
3Y Dump truck model Y
No computer

Opportunity Cost of Capital = 14%

b) Borrow $10,000. choose Projects:

1 Repair bulldozer
2B Backhoe model B
3Y Dump truck model Y
4K Computer model K

15-4

P&J Brewing has the following capital structure:

Type	Amount	Average Minimum Return
Mortgages	$25,000,000	7%
Bonds	180,000,000	9%
Common Stock	100,000,000	10%
Preferred Stock	50,000,000	8%
Retained Earnings	120,000,000	10%

Determine the WACC for P&J.

Solution

$$\text{WACC} = \frac{25M(.07) + 180M(.09) + 100M(.10) + 50M(.08) + 120M(.10)}{(25M + 180M + 100M + 50M + 120M)} = 9.25\%$$

Chapter 16

Economic Analysis in The Public-Sector

16-1

A city engineer is considering installing an irrigation system. He is trying to decide which one of two alternatives to select. The two alternatives have the following cash flow:

Year	A	B
0	-$15,000	-$25,000
1-10	+5,310	+7,900

If interest is 12%, which alternative should the engineer select? Assume no salvage value. Use incremental benefit/cost analysis.

Solution

	A	B
PW_{COST}	-15,000	-25,000
$PW_{BENEFITS}$	5,310(P/A, 12%, 10) = 30,000	7,900(P/A, 12%, 10) = 44,635
B/C	30,000/15,000 = 2	44,635/25,000 = 1.79

B-A

PW_{COST} -10,000

ΔAnnual Benefit = +2,590

$PW_{BENEFITS}$ 2,590(P/A, 12%, 10) = 14,634

ΔB/C = 14,634/10,000 = 1.46

1.46 > 1 Therefore choose B, higher cost alternative

16-2

With interest at 10 %, what is the benefit-cost ratio for this government project?

Initial cost	\$2,000,000
Additional costs at end of years 1 & 2	30,000
Benefits at end of years 1 & 2	0
Annual benefits at end of years 3 - 10	90,000

Solution

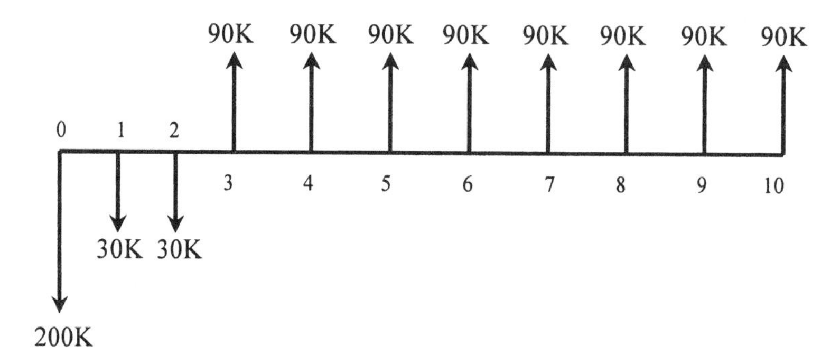

PW Cost = 200K + 30K(P/A, 10%, 2) = 252,080

PW Benefits = 90K(P/A, 10%, 8)(P/F, 10%, 2) = 396,800.

B/C = 396,800/252,080
= 1.574

16-3

A van is bought for \$15,000 in Year 0. It generates revenues of \$25,000 per year and expenses of \$20,000 per year in each of the Years 1 thru 5. Its salvage value is zero after 5 years. Ignoring income taxes, compute the benefit/cost ratio if i = 10%.

Solution

$PW_B = 25{,}000(P/A, 10\%, 5) = \$94{,}775$

$PW_C = 20{,}000\ (P/A, 10\%, 5) + 15{,}000 = \$90{,}820$

$$\frac{B}{C}\text{Ratio} = \frac{PW_B}{PW_C} = \frac{94{,}775}{90{,}820} = 1.0435$$

16-4

A tax-exempt municipality is considering the construction of an impoundment for city water supplies. Two different sites have been selected as technically, politically, socially, and financially feasible. The city council has asked you to do a benefit-cost analysis of the alternatives and recommend the best site. The city uses a 6% interest rate in all analysis of this type.

Year	Rattlesnake Canyon	Blue Basin
0	-$15,000,000	-$27,000,000
1 - 75	+2,000,000	+3,000,000

Which should you recommend?

Solution

		Rattlesnake	Blue Basin
$\frac{\text{PW of Benefits}}{\text{PW of Cost}}$	=	$\frac{2 \times 10^6(P/A, 6\%, 75)}{15 \times 10^6}$	$\frac{3 \times 10^6(P/A, 6\%, 75)}{27 \times 10^6}$
$\frac{B}{C}$ ratio	=	2.19 >1 (OK)	1.83 > 1 (OK)

Incremental Analysis

Year	BB - RC
0	-12,000,000
1 - 75	+1,000,000

$$\frac{B}{C} = \frac{1 \times 10^6(P/A, 6\%, 75)}{12 \times 10^6} = 1.37 > 1$$

∴ choose higher cost alternative, Blue Basin

16-5

A city engineer is deciding which of two bids for a new computer to accept. Using benefit-cost analysis, which alternative should be selected if the interest rate is 10% per year?

Computer	Cost	Annual Benefits	Salvage	Useful life
A	$48,000	$13,000	$0	6 years
B	40,000	12,000	0	6 years

Solution

Alternative A:
$PW_B = 13{,}000(P/A, 10\%, 6) = \$56{,}615.$
$PW_C = 48{,}000$
$B/C = 56{,}615/48{,}000 = 1.179$ (OK)

Alternative B
$PW_B = 12{,}000(P/A, 10\%, 6) = \$52{,}260$
$PW_C = 40{,}000$
$B/C = 52{,}260 / 40{,}000 = 1.3065$ (OK)

Incremental Analysis

$$\frac{\Delta B}{\Delta C} = \frac{56{,}516 - 52{,}260}{48{,}000 - 40{,}000} = \frac{4{,}355}{8{,}000} = .544$$

$$\frac{\Delta B}{\Delta C} < 1.0 \quad \therefore \text{select least cost alternative, B}$$

16-6

Given four mutually exclusive alternatives, each with a useful life of 20 years and no salvage value, have been presented to the city council of Anytown USA. Which alternative should be selected?

	A	B	C	D
PW of Costs	\$4,000	\$9,000	\$6,000	\$2,000
PW of Benefits	6,480	11,250	5,700	4,700

Solution

	A	B	C	D
B/C	1.62	2.35	0.95	2.35

$C < 1 \therefore$ eliminate

(rearrange)	D	A	B
PW of Cost	2,000	4,000	9,000
PW of Benefits	4,700	6,480	11,250

	A - D	B - A		
ΔB	1,780	4,770		
ΔC	2,000	5,000		
ΔB/ΔC	.89.	95	≤ 1.0	∴ choose D least expensive

16-7

The state engineer estimates that the cost of a canal is $680 million. The legislative analyst estimates the equivalent annual cost of the investment for the canal to be $20.4 million. If the analyst expects the canal to last indefinitely, what interest rate is he using to compute the equivalent annual cost (EAC)?

If the canal lasts only 50 years, what interest rate will the analyst be assuming if he believes the EAC to be the same $20.4 million?

Solution

(a) $A = P(A/P, i\%, n)$

For $n \rightarrow \infty$, $(A/P, i\%, n) = i$

$A = P \times i$

$i = A/P = \frac{20.4}{680} = .03$ or $i = 3\%$

(b) $A = P\ (A/P, i\%, n)$

$A = P\ (A/P, i\%, 50)$

$(A/P, i\%, 50) = \frac{20.4}{680} = .03$ $i = 1.75\%$

16-8

The local city recorder is deciding between two different phone answering systems for her office. The information concerning the two machines is presented below.

Machine	Cost	Annual Savings	Useful Life	Salvage
X	$1,000	$300	5 years	$0
Y	1,200	325	5 years	0

With an assumed interest rate of 12%, which system would you recommend?

Solution

Machine X: $PW_C = \$1,000$

$PW_B = \$300(P/A, 12\%, 5) = \$1,081.5$

B/C ratio = 1.0815/1,000 = 1.0815

Machine Y: $PW_C = \$1,200$

$PW_B = \$325(P/A, 12\%, 5) = \$1,171.63$

B/C ratio = 1,171.63/1,200 = .9763 $\therefore$ Select machine X

16-9

The city council of Arson, Michigan is debating whether to buy a new fire truck to increase protection for the city. The financial analyst has prepared the following data:

	Truck A	Truck B
First cost	$50,000	$60,000
Annual Maintenance	5,000	4,000
Useful Life	7 years	7 years
Salvage value	6,000	6,000
Annual Reduction in fire damage	20,000	21,000

(a) Using the modified B/C ratio method determine whether the city should buy a new truck, and which one to buy if it will be paid for with money borrowed at an interest rate 7%.
(b) How would the decision be affected if inflation were considered? Assume maintenance cost, salvage and fire damage are responsive to inflation.
(c) Is it possible would change if the interest rate were lower? Why or why not?

Solution

(a) In the modified B/C ratio, all annual cash flows belong in the numerator while first cost and salvage belong in the denominator. Either present or uniform equivalent methods may be used to relate cash flows.

$$\left(\frac{B}{C}\right)_A = \frac{20{,}000 - 5{,}000}{50{,}000(A/P, 7\%, 7) - 5{,}000(A/F, 7\%, 7)} = 1.72\,(>1) \qquad \therefore \text{A is acceptable}$$

$$\left(\frac{B}{C}\right)_{B\text{-}A} = \frac{(21{,}000 - 20{,}000) - (4{,}000 - 5{,}000)}{10{,}000(A/P, 7\%, 7)} = 1.08\,(>1) \qquad \therefore \text{B is better than A}$$

Truck B should be purchased

(b) Since both future cost (maintenance) and benefits (reduced damage and salvage) are responsive to inflation, the decisions are not affected by inflation.

(c) No. Lower i (MARR) favors higher investment cost projects and truck B is already the highest cost alternative.

16-10

The Tennessee department of Transportation (TDOT) is considering building its first "tolled-bypass" around the town of Greenfield, Tennessee. The initial cost of the bypass is estimated to be $5.7 million. The installation of the tollbooth along the bypass is estimated to be about $1.3 million. The annual maintenance cost of the bypass is to be $85,000 each year while the annual maintenance cost of the tollbooth is to be $65,000 each year. Due to the construction of the bypass, Greenfield is expected to receive estimated tax savings of $225,000 per year. In addition, TDOT has projected savings of $100,000 each year. Once the tollbooth is constructed, each car that passes through the toll is expected to pay a fare of $0.90. The city estimates that there will be a total of 500,000 cars using the bypass each year. Other relevant data is listed below.

Resurfacing cost (every 7 years)	4% of bypass initial cost
Shoulder grading/rework	90% of resurfacing cost

If the state uses an interest rate of 7%, should the "toll-bypass" be constructed? What is the project's total yearly worth? Assume perpetual life.

Solution:

A. Use benefit/cost analysis. (Use AW analysis)

Benefits	
Greenfield Tax Annual Revenue	$225,000.00
Bypass Construction Annual Savings	$100,000.00
Tollbooth Revenue $0.90(500,000)	$450,000.00
	AW_B = $ 775,000.00

Costs	
Toll First Cost $1,300,000(A/P, 7%, ∞)	$ 91,000.00
Bypass First Cost $5,700,000(A/P, 7%, ∞)	$399,000.00
Toll Maintenance	$ 85,000.00
Bypass Maintenance	$ 65,000.00
Resurfacing 0.04($5,700,000(A/P, 7%, 7))	$ 26,536.80
Shoulder Grading/Rework 0.90(0.04($5,700,000(A/P, 7%, 7)))	$ 23,721.12
	AW_C = ($690,257.92)

$$\frac{B}{C} = \frac{AW_B}{AW_C} = \frac{\$775{,}000.00}{\$690{,}257.92} = 1.123$$

- The bypass should be constructed

B. Total Annual Worth

$$AW_T = AW_B + AW_C = \$\ 775{,}000.00 + (\$\ 690{,}257.92) = \$84{,}742.08$$

16-11

The town of Oakville is evaluating a proposal that it erect and operate a structure for parking in the downtown area. Numerous design proposals were considered. Data for the two best proposals is presented below.

	Design A	Design B
Cost of site and construction	\$2,220,000	\$1,850,000
Annual revenue from parking fees	765,000	575,000
Annual maintenance cost	410,000	295,000
Service life	20 years	20 years

At the end of each five year period the parking facility will require a maintenance overhaul. The cost for design A is estimated to be \$650,000 and for design B, \$375,000.

At the end of the service life the facility will be torn down and the land sold. Demolition costs and proceeds from the sale of the land are presented below.

Demolition costs	\$530,000	\$550,000
Proceeds from sale of land	530,000	530,000

If the city's interest rate is 10%, determine the B/C ratio of each design proposal.

Solution

Design A

PW_B = 765,000(P/A, 10%, 20) = \$6,513,210

PW_C = 2,220,000 + 410,000(P/A, 10%, 20) + 650,000(P/F, 10%, 5,10, and 15) = \$6,500,510

$$\frac{B}{C} = \frac{\$6,513,210}{\$6,500,510} = 1.001$$

(Note: Demolition and proceeds from sale net to \$0)

Design B

PW_B = 575,000 + 530,000(P/F, 10%, 20) = \$4,974,308

PW_C = 1,850,000 + 295,000(P/A, 10%, 20) + 375,000(P/F, 10%, 5, 10, and 15)
+ 550,000(P/F, 10%, 20) = \$4,910,535

$$\frac{B}{C} = \frac{\$4,974,308}{\$4,910,535} = 1.013$$

16-12

A new Water Treatment Plant will cost the city of Frogjump $2,000,000 dollars to build and $150,000 per year to operate for its 30-year life. At the end of 30 yrs, the salvage value will be 0. Due to more efficient operation of the water plant, it is expected to lower the cost of utility bills for each customer $50 per year. There are 6,000 customers that are billed in Frogjump. The plant will reduce air quality by an estimated $5 per resident per year. The population of Frogjump is currently 18,000 and is expected to remain relatively constant over the life of the plant. If 3% is used for the evaluation of public works projects, should the Water Treatment Plant be built?

Solution

First Cost: $2,000,000
Annual Operating Cost: $150,000/yr
Additional Annual Cost: $5/year x 18,000 residents = $90,000/year
Annual Benefits: $50/year x 6,000 customers = $300,000/year
Life: 30 yrs
Interest: 3%

Using B/C ratio AW_B/AW_C should be greater than or equal to 1

AW_B = $300,000

AW_C = $2,000,000(A/P, 3%, 30) = $342,000

$$\frac{B}{C} = \frac{AW_B}{AW_C} = \frac{\$300{,}000}{\$342{,}000} = 0.8772$$

- New Water Treatment Plant should not be built

16-13

The expansion of the hotel and conference center at Wicker Valley State Park is under study. The initial investment and annual operating benefits and cost are very different due to the differing magnitudes of the project under consideration. These can be summarized as follows:

	Alternative A	Alternative B	Alternative C
Investment cost	$180,000	$100,000	$280,000
Annual operating costs	16,000	12,000	28,000
Annual benefits	53,000	35,000	77,000

Using a MARR of 10%, which alternative should be selected for a ten-year useful life analysis period? Use benefit/cost ratio analysis.

Solution

1. B/C ratios of individual alternatives.

$$\left(\frac{B}{C}\right)_A = \frac{AW_B}{AW_C} = \frac{53{,}000}{180{,}000(A/P, 10\%, 10) + 16{,}000} = 1.17(>1)$$

$$\left(\frac{B}{C}\right)_B = \frac{AW_B}{AW_C} = \frac{35{,}000}{100{,}000(A/P, 10\%, 10) + 12{,}000} = 1.24(>1)$$

$$\left(\frac{B}{C}\right)_C = \frac{AW_B}{AW_C} = \frac{77{,}000}{280{,}000(A/P, 10\%, 10) + 28{,}000} = 1.05(>1)$$

$\therefore$ All economically attractive

2. Incremental B/C Analysis

Cost B < Cost A < Cost C

<u>A - B</u>
ΔBenefits = 53,000 - 35,000 = 18,000
ΔCosts = (180,000 - 1000,000)(A/P, 10%, 10) + (16,000 - 12,000) = 17,016

$$\left(\frac{B}{C}\right)_{A\text{-}B} = \frac{18{,}000}{17{,}016} = 1.06(>1) \qquad \therefore \text{Keep A}$$

<u>C - A</u>
ΔBenefits = 77,000 - 53,000 = 24,000
ΔCosts = (280,000 - 180,000)(A/P, 10%, 10) + (28,000 - 16,000) = 28,270

$$\left(\frac{B}{C}\right)_{C\text{-}A} = \frac{24{,}000}{28{,}270} = 0.85(<1) \qquad \therefore \text{Keep A}$$

Select Alternative A. (If we consider the operating costs as a reduction to the annual benefits a different numerical value for the B/C ratio might be found. The decision of which alternative is best will, however, be the same.)

16-14

Using benefit-cost ratio analysis, determine which one of the following alternatives should be selected. Each alternative has six-year useful life. Assume a 10% MARR.

	A	B	C	D
First cost	\$880	\$560	\$700	\$900
Annual benefit	240	130	110	250
Annual cost	80	20	0	40
Salvage value	300	200	440	110

Solution

<u>A</u>

$AW_{BENEFITS} = 240 + 300(A/F, 10\%, 6) = 278$

$AW_{COSTS} = 880(A/P, 10\%, 6) + 80 = 282$

$B/C = .98$ (< 1, so, eliminate)

<u>B</u>

$AW_{BENEFITS} = 130 + 200(A/F, 10\%, 6) = 155.9$

$AW_{COSTS} = 560(A/P, 10\%, 6) + 20 = 148.57$

$B/C = 1.04$

<u>C</u>

$AW_{BENEFITS} = 110 + 440(A/F, 10\%, 6) = 167$

$AW_{COSTS} = 700(A/P, 10\%, 6) + 0 = 160$

$B/C = 1.04$

<u>D</u>

$AW_{BENEFITS} = 250 + 110(A/F, 10\%, 6) = 264$

$AW_{COSTS} = 900(A/P, 10\%, 6) + 40 = 246$

$B/C = 1.07$

<u>D-C</u>

$AW_{BENEFITS} = (250 - 110) + (110 - 440)(A/F, 10\%, 6) = 97$

$AW_{COSTS} = (900 - 700)(A/P, 10\%, 6) + (40 - 0) = 85$

$B/C = 1.14$ → choose D

<u>(D)-(B)</u>

$AW_{BENEFITS} = (250 - 130) + (110-200)(A/F, 10\%, 6) = 108$

$AW_{COSTS} = (900 - 560)(A/P, 10\%, 6) + (40 - 20) = 98$

$B/C = 1.10$ → choose D

16-15

A new electric generation plant is expected to cost \$43,250,000 to complete. The revenues generated by the new plant are expected to be \$3,875,000 per year while operational expenses are estimated to \$2,000,000 per year. If the plant is expected to last forty years and the electric authority uses 3% as its cost of capital, determine if the plant should be built? (Use B/C ratio analysis.)

Solution

$$AW_{BENEFITS} = \$3,875,000$$
$$AW_{COSTS} = 43,500,000(A/P, 3\%, 40) + 2,000,000 = \$3,872,725$$

$$B/C = \frac{AW_{BENEFITS}}{AW_{COSTS}} = \frac{3,875,000}{3,872,725} = 1 \qquad \text{Build Plant}$$

Chapter 17

Rationing Capital Among Competing Projects

17-1
The following independent and indivisible investment opportunities are available:

Investment	Initial Cost	Rate of Return
A	$200	20%
B	100	22%
C	50	19%
D	100	18%
E	50	15%
F	Unlimited	7%

(a) Which investment(s) should be selected if the minimum attractive rate of return (MARR) is greater then or equal to 18% assuming an unlimited budget?
(b) Which investment(s) should be selected if the available budget is $400 and the MARR is greater than or equal to 14%?

Solution

(a) A, B, C, D → Choose all IRR's ≥ 18% since budget is unlimited.

(b) A, B, D → Choose D instead of C because it yields a greater overall return and fully invests the $400 budget.

Investment	Initial Cost	Return$	Return in $
A	$200	20%	$40.00
B	100	22%	22.00
C	50	19%	9.50
D	100	18%	18.00

Total $ return on A + B + C = 40 + 22 + 9.50 +50(.14)* = $78.50
Total $ return on A + B + D = 40 + 22 +18 = $80.00

*Assumes the remaining $50 can be invested at the MARR. This is not always true. Thus yielding an even lower $ return.

17-2

A city engineer calculated the present worth of benefits and costs of a number of possible projects, based on 10% interest and a 10 year analysis period.

Costs and Benefits in 1000's

Project:	A	B	C	D	E	F	G
Present Worth of Costs	75	70	50	35	60	25	70
Present Worth of Benefits	105	95	63	55	76	32	100

If 10% is a satisfactory minimum attractive rate of return (MARR), which project(s) should be selected if $180,000 is available for expenditure?

Solution

Project:	A	B	C	D	E	F	G
Present Worth of Costs	75	70	50	35	60	25	70
Present Worth of Benefits	105	95	63	55	76	32	100
NPW	30	25	13	20	16	7	30
NPW/C	.400	.357	.260	.571	.267	.280	.428
Rank	3	4	7	1	6	5	2

D + G + A = 35 K + 70 K + 75 K = 180 K

Choose projects D, G, A

17-3

Barber Brewing is in the process of determining the capital budget for the coming year. The following projects are under consideration.

	A	B	C	D
First Cost	$10,000	$13,000	$20,000	$33,000
Annual Income	10,000	9,078	16,000	16,455
Annual Cost	7,362	5,200	11,252	7,300

All projects have a five year useful life. Which alternative(s) should be selected if Barber's budget is set at $50,000?

Solution

NPW = 0 at IRR.

0 = -First cost + net income (P/A, i%, n)   Therefore (P/A, i%, n) = First cost/net income

			Rank
IRR_A	P/A = 10,000/2,638 = 3.791	IRR = 10%	3
IRR_B	P/A = 13,000/3,878 = 3.352	IRR = 15%	1
IRR_C	P/A = 20,000/4,748 = 4.212	IRR = 6%	4
IRR_D	P/A = 33,000/9,155 = 3.605	IRR = 12%	2

Choose projects B and D. Total = 13,000 + 33,000 = $46,000

17-4

ABC Builders has asked each of its four regional managers to submit requests for capital outlays for the following fiscal year. The CEO of ABC has decided to fund the top request from each region and to fund two additional requests with the provision that no region has more than two projects funded. Using the information provided below determine which projects should be funded.

Region	Project	Cost	Annual Benefit	Life (Years)
Southern (S)	A	$ 90,000	$16,400	15
	B	40,000	15,000	5
	C	60,000	20,400	5
	D	120,000	27,600	20
Midwest (MW)	A	50,000	10,000	20
	B	120,000	36,700	15
	C	75,000	21,600	5
	D	50,000	16,200	5
New England (NE)	A	50,000	16,700	20
	B	80,000	23,500	5
	C	75,000	26,100	10
Pacific (P)	A	60,000	16,900	15
	B	50,000	15,300	10

Solution

Region	Project	Cost	IRR
Southern (S)	A	$ 90,000	16.3%
	B	40,000	25.4
	C	60,000	20.8
	D	120,000	22.6
Midwest (MW)	A	50,000	19.4
	B	120,000	30.0
	C	75,000	13.5
	D	50,000	18.6
New England (NE)	A	50,000	33.3
	B	80,000	14.4
	C	75,000	32.8
Pacific (P)	A	60,000	27.4
	B	50,000	28.0

Choose (S)B, (MW)B, (NE)A, (P)B, (NE)C, and (P)A

Capital Budget = 40,000 + 120,000 + 50,000 + 50,000 + 75,000 + 60,000
= $395,000

Chapter 18

Accounting and Engineering Economy

18-1

The following information is has been taken from the financial statements available for the ABC Company:

Accounts payable	$ 4,000
Accounts receivable	12,000
Income taxes	6,000
Owner's equity	75,000
Cost of goods sold	42,000
Selling expense	10,000
Sales revenue	80,000

Determine the net income.

Solution

Net income = Sales revenue - Cost of goods sold - Selling expense - Income taxes
= 80,000 - 42,000 - 10,000 - 6,000
= $22,000

18-2

Billy Bob's Towing and Repair Service has provided the following financial information:

Cash	$80,000
Accounts receivable	120,000
Accounts payable	200,000
Securities	75,000
Parts Inventories	42,000
Prepaid expenses	30,000
Accrued expense	15,000

Determine the (a) current ratio, (b) the quick ratio, and the (c) available working capital.

Solution

(a) $\text{Current ratio} = \dfrac{\text{Current Assets}}{\text{Current Liabilities}} = \dfrac{80{,}000 + 120{,}000 + 75{,}000 + 42{,}000}{200{,}000 + 15{,}000} = 1.47$

(b) $\text{Quick ratio} = \dfrac{\text{Current Assets - Inventories}}{\text{Current Liabilities}} = \dfrac{80{,}000 + 120{,}000 + 75{,}000}{200{,}000 + 15{,}000} = 1.28$

(c) Working capital = Current assets - Current liabilities = $102,000

18-3

The following financial information was taken from the income statement of Firerock Industries

Revenues	
Sales	$3,200,000
Operating revenue	2,000,000
Nonoperationg revenue	3,400,000
Expenses	
Total operating expenses	6,700,000
Interest payments	500,000

Taxes paid for the year equaled $110,000. Determine the (a) net income before taxes, (b) net profit (loss), (c) interest coverage and (d) net profit ratio.

Solution

(a) Net income before taxes = 3.2M + 2.0M + 3.4M - 6.7M - .5M = $1,400,000

(b) Net profit = 1.4M - .11M = $1,290,000

(c) $\text{Interest coverage} = \dfrac{\text{Total Income}}{\text{Interest Payments}} = \dfrac{1.9\text{M}}{.5\text{M}} = 3.8$

(d) $\text{Net profit ratio} = \dfrac{\text{Net Profit}}{\text{Net Sales Revenue}} = \dfrac{1.29\text{M}}{3.2\text{M}} = .4031$

18-4

Determine value of the retained earnings for Lavelle Manufacturing.

Current liabilities	$4,000,000
Current assets	6,500,000
Fixed assets	4,000,000
Common stock	2,500,000
Long-term liabilities	2,000,000
Preferred stock	500,000
Other assets	1,500,000
Capital surplus	1,000,000

Solution

Equity = Common stock + Preferred stock + Capital surplus + Retained earnings

Equity = 2.5M + .5M + 1M + Retained earnings
= 4M + Retained earnings

Assets = Liabilities + Equity

6.5M + 4M + 1.5M = 4M + 2M + 4M + Retained earnings

Retained earnings = $2,000,000

18-5

Abby Manufacturing produces numerous children's toys. The Dr. Dolittle Farm is one of the most popular sellers. Indirect cost to be allocated to production of the toy is to be calculated based on direct materials allocation. The total production overhead for the facility where the toy is produced is $750,000. The direct material total for the facility is $8,350,000.

The cost of the direct materials used in production of the Dr. Dolittle is $7.45 per unit. The total labor (both direct and indirect) for the production of the toy is $9.35 per unit. The production schedule for the coming year calls for 300,000 units to be produced. If Abby desires a 35% profit on the toy, what should the wholesale price be?

Solution

Total labor = 9.35 x 300,000 = 2,805,000
Total materials = 7.45 x 300,000 = 2,235,000
Overhead = (2,235,000/8,350,000) x 750,000 = $200,748.50

Total production cost = 2,805,000 + 2,235,000 + 200,748.50 = $5,240,748.50
Cost per unit = 5,240,748.50/300,000 = $17.47

Wholesale price = 17.47(1.35) = $23.58

18-6

Brown Box Inc manufactures shipping boxes for a wide variety of industries. Their model XLLong has the following direct manufacturing costs per unit:

Direct materials costs	$0.25
Direct labor costs	2.75

Overhead for the entire manufacturing plant is $4,000,000 per year. Direct labor costs are used to allocate the overhead. The total direct labor costs are estimated to be $5,500,000. The expected demand for this particular model is 200,000 boxes for the year. Determine the cost per unit.

Solution

Cost per unit = Direct materials cost + Direct labor costs + Overhead costs

Overhead costs

Direct labor cost = 200,000 x $2.75 = 550,000

$$\text{Allocation of overhead} = 4{,}000{,}000 \times \frac{550{,}000}{5{,}500{,}000} = 400{,}000/200{,}000 = \$2/\text{box}$$

Cost per unit = .25 + 2.75 + 2.00 = $5.00

18-7

The following financial information is know about Rapid Delivery Inc.:

Acid-test ratio	1.3867
Cash on hand	$ 72,000
Accounts receivable	102,000
Market value of securities held	34,000
Inventories	143,000
Other assets	16,000
Fixed assets	215,000
Total liabilities	400,000

Determine the (a) current assets, (b) current liabilities, (c) total assets, and (d) owner's equity.

Solution

(a) Current assets = 72,000 + 102,000 + 34,000 + 143,000 = $351,000

(b) Acid-rest ratio $= \frac{\text{Current Assets - Inventories}}{\text{Current Liabilities}}$

Current liabilities $= \frac{\text{Current Assets - Inventories}}{\text{Acid-test ratio}} = \frac{72{,}000 + 102{,}000 + 34{,}000}{1.3867} = \$149{,}996$

(c) Total assets = 351,000 + 215,000 + 16,000 = $582,000

(d) Owner's equity = Total assets - Total liabilities
= 582,000 - 400,000
= $182,000

18-8

Determine the current and quick ratios for Harbor Master Boats Inc. Does the company appear to be reasonably sound from a financial viewpoint based on these two ratios?

Harbor Master Boats Inc.
Balance Sheet, January 1, 200X

Assets		**Liabilities**	
Current Assets		Current Liabilities	
Cash	900,000	Accounts Payable	2,400,000
Accounts Receivable	1,100,000	Notes Payable	2,000,000
Inventory	2,000,000	Accrued Expense	900,000
Total Current Assets	4,000,000	Total Current Liabilities	5,300,000
Fixed Assets		Long Term Debt	3,000,000
Land	300,000	Total Liabilities	8,300,000
Plant	2,500,000	**Equity**	
Equipment	6,000,000	Stock	2,000,000
Total Fixed Assets	8,800,000	Retained Earnings	2,500,000
		Total Net Worth	4,500,000
Total Assets	12,800,000	Total Liabilities and Net Worth	12,800,000

Solution

$$\text{Current ratio} = \frac{\text{Current Assets}}{\text{Current Liabilities}} = \frac{4{,}000{,}000}{5{,}300{,}000} = .755$$

$$\text{Quick ratio} = \frac{\text{Current Assets - Inventories}}{\text{Current Liabilities}} = \frac{4{,}000{,}000 - 2{,}000{,}000}{5{,}300{,}000} = .377$$

Based on these two ratios the company is not in very sound financially. The current typically should be greater than 2. The quick ratio indicates the company's inability to pay off current liabilities with "quick" capital.

APPENDIX B

Compound Interest Tables

Values of Interest Factors When N Equals Infinity

Single Payment:

$(F/P, i, \infty) = \infty$

$(P/F, i, \infty) = 0$

Arithmetic Gradient Series:

$(A/G, i, \infty) = 1/i$

$(P/G, i, \infty) = 1/i^2$

Uniform Payment Series:

$(A/F, i, \infty) = 0$

$(A/P, i, \infty) = i$

$(F/A, i, \infty) = \infty$

$(P/A, i, \infty) = 1/i$

1/4% **Compound Interest Factors** 1/4%

	Single Payment		Uniform Payment Series				Arithmetic Gradient		
	Compound Amount Factor	Present Worth Factor	Sinking Fund Factor	Capital Recovery Factor	Compound Amount Factor	Present Worth Factor	Gradient Uniform Series	Gradient Present Worth	
n	Find *F* Given *P* *F/P*	Find *P* Given *F* *P/F*	Find *A* Given *F* *A/F*	Find *A* Given *P* *A/P*	Find *F* Given *A* *F/A*	Find *P* Given *A* *P/A*	Find *A* Given *G* *A/G*	Find *P* Given *G* *P/G*	*n*
1	1.003	.9975	1.0000	1.0025	1.000	0.998	0	0	**1**
2	1.005	.9950	.4994	.5019	2.003	1.993	0.504	1.005	**2**
3	1.008	.9925	.3325	.3350	3.008	2.985	1.005	2.999	**3**
4	1.010	.9901	.2491	.2516	4.015	3.975	1.501	5.966	**4**
5	1.013	.9876	.1990	.2015	5.025	4.963	1.998	9.916	**5**
6	1.015	.9851	.1656	.1681	6.038	5.948	2.498	14.861	**6**
7	1.018	.9827	.1418	.1443	7.053	6.931	2.995	20.755	**7**
8	1.020	.9802	.1239	.1264	8.070	7.911	3.490	27.611	**8**
9	1.023	.9778	.1100	.1125	9.091	8.889	3.987	35.440	**9**
10	1.025	.9753	.0989	.1014	10.113	9.864	4.483	44.216	**10**
11	1.028	.9729	.0898	.0923	11.139	10.837	4.978	53.950	**11**
12	1.030	.9705	.0822	.0847	12.167	11.807	5.474	64.634	**12**
13	1.033	.9681	.0758	.0783	13.197	12.775	5.968	76.244	**13**
14	1.036	.9656	.0703	.0728	14.230	13.741	6.464	88.826	**14**
15	1.038	.9632	.0655	.0680	15.266	14.704	6.957	102.301	**15**
16	1.041	.9608	.0613	.0638	16.304	15.665	7.451	116.716	**16**
17	1.043	.9584	.0577	.0602	17.344	16.624	7.944	132.063	**17**
18	1.046	.9561	.0544	.0569	18.388	17.580	8.437	148.319	**18**
19	1.049	.9537	.0515	.0540	19.434	18.533	8.929	165.492	**19**
20	1.051	.9513	.0488	.0513	20.482	19.485	9.421	183.559	**20**
21	1.054	.9489	.0464	.0489	21.534	20.434	9.912	202.531	**21**
22	1.056	.9465	.0443	.0468	22.587	21.380	10.404	222.435	**22**
23	1.059	.9442	.0423	.0448	23.644	22.324	10.894	243.212	**23**
24	1.062	.9418	.0405	.0430	24.703	23.266	11.384	264.854	**24**
25	1.064	.9395	.0388	.0413	25.765	24.206	11.874	287.407	**25**
26	1.067	.9371	.0373	.0398	26.829	25.143	12.363	310.848	**26**
27	1.070	.9348	.0358	.0383	27.896	26.078	12.852	335.150	**27**
28	1.072	.9325	.0345	.0370	28.966	27.010	13.341	360.343	**28**
29	1.075	.9301	.0333	.0358	30.038	27.940	13.828	386.366	**29**
30	1.078	.9278	.0321	.0346	31.114	28.868	14.317	413.302	**30**
36	1.094	.9140	.0266	.0291	37.621	34.387	17.234	592.632	**36**
40	1.105	.9049	.0238	.0263	42.014	38.020	19.171	728.882	**40**
48	1.127	.8871	.0196	.0221	50.932	45.179	23.025	1 040.22	**48**
50	1.133	.8826	.0188	.0213	53.189	46.947	23.984	1 125.96	**50**
52	1.139	.8782	.0180	.0205	55.458	48.705	24.941	1 214.76	**52**
60	1.162	.8609	.0155	.0180	64.647	55.653	28.755	1 600.31	**60**
70	1.191	.8396	.0131	.0156	76.395	64.144	33.485	2 147.87	**70**
72	1.197	.8355	.0127	.0152	78.780	65.817	34.426	2 265.81	**72**
80	1.221	.8189	.0113	.0138	88.440	72.427	38.173	2 764.74	**80**
84	1.233	.8108	.0107	.0132	93.343	75.682	40.037	3 030.06	**84**
90	1.252	.7987	.00992	.0124	100.789	80.504	42.820	3 447.19	**90**
96	1.271	.7869	.00923	.0117	108.349	85.255	45.588	3 886.62	**96**
100	1.284	.7790	.00881	.0113	113.451	88.383	47.425	4 191.60	**100**
104	1.297	.7713	.00843	.0109	118.605	91.480	49.256	4 505.93	**104**
120	1.349	.7411	.00716	.00966	139.743	103.563	56.512	5 852.52	**120**
240	1.821	.5492	.00305	.00555	328.306	180.312	107.590	19 399.75	**240**
360	2.457	.4070	.00172	.00422	582.745	237.191	152.894	36 264.96	**360**
480	3.315	.3016	.00108	.00358	926.074	279.343	192.673	53 821.93	**480**

1/2% Compound Interest Factors 1/2%

	Single Payment		Uniform Payment Series				Arithmetic Gradient		
	Compound Amount Factor	Present Worth Factor	Sinking Fund Factor	Capital Recovery Factor	Compound Amount Factor	Present Worth Factor	Gradient Uniform Series	Gradient Present Worth	
	Find F Given P	Find P Given F	Find A Given F	Find A Given P	Find F Given A	Find P Given A	Find A Given G	Find P Given G	
n	F/P	P/F	A/F	A/P	F/A	P/A	A/G	P/G	n
1	1.005	.9950	1.0000	1.0050	1.000	0.995	0	0	1
2	1.010	.9901	.4988	.5038	2.005	1.985	0.499	0.991	2
3	1.015	.9851	.3317	.3367	3.015	2.970	0.996	2.959	3
4	1.020	.9802	.2481	.2531	4.030	3.951	1.494	5.903	4
5	1.025	.9754	.1980	.2030	5.050	4.926	1.990	9.803	5
6	1.030	.9705	.1646	.1696	6.076	5.896	2.486	14.660	6
7	1.036	.9657	.1407	.1457	7.106	6.862	2.980	20.448	7
8	1.041	.9609	.1228	.1278	8.141	7.823	3.474	27.178	8
9	1.046	.9561	.1089	.1139	9.182	8.779	3.967	34.825	9
10	1.051	.9513	.0978	.1028	10.228	9.730	4.459	43.389	10
11	1.056	.9466	.0887	.0937	11.279	10.677	4.950	52.855	11
12	1.062	.9419	.0811	.0861	12.336	11.619	5.441	63.218	12
13	1.067	.9372	.0746	.0796	13.397	12.556	5.931	74.465	13
14	1.072	.9326	.0691	.0741	14.464	13.489	6.419	86.590	14
15	1.078	.9279	.0644	.0694	15.537	14.417	6.907	99.574	15
16	1.083	.9233	.0602	.0652	16.614	15.340	7.394	113.427	16
17	1.088	.9187	.0565	.0615	17.697	16.259	7.880	128.125	17
18	1.094	.9141	.0532	.0582	18.786	17.173	8.366	143.668	18
19	1.099	.9096	.0503	.0553	19.880	18.082	8.850	160.037	19
20	1.105	.9051	.0477	.0527	20.979	18.987	9.334	177.237	20
21	1.110	.9006	.0453	.0503	22.084	19.888	9.817	195.245	21
22	1.116	.8961	.0431	.0481	23.194	20.784	10.300	214.070	22
23	1.122	.8916	.0411	.0461	24.310	21.676	10.781	233.680	23
24	1.127	.8872	.0393	.0443	25.432	22.563	11.261	254.088	24
25	1.133	.8828	.0377	.0427	26.559	23.446	11.741	275.273	25
26	1.138	.8784	.0361	.0411	27.692	24.324	12.220	297.233	26
27	1.144	.8740	.0347	.0397	28.830	25.198	12.698	319.955	27
28	1.150	.8697	.0334	.0384	29.975	26.068	13.175	343.439	28
29	1.156	.8653	.0321	.0371	31.124	26.933	13.651	367.672	29
30	1.161	.8610	.0310	.0360	32.280	27.794	14.127	392.640	30
36	1.197	.8356	.0254	.0304	39.336	32.871	16.962	557.564	36
40	1.221	.8191	.0226	.0276	44.159	36.172	18.836	681.341	40
48	1.270	.7871	.0185	.0235	54.098	42.580	22.544	959.928	48
50	1.283	.7793	.0177	.0227	56.645	44.143	23.463	1 035.70	50
52	1.296	.7716	.0169	.0219	59.218	45.690	24.378	1 113.82	52
60	1.349	.7414	.0143	.0193	69.770	51.726	28.007	1 448.65	60
70	1.418	.7053	.0120	.0170	83.566	58.939	32.468	1 913.65	70
72	1.432	.6983	.0116	.0166	86.409	60.340	33.351	2 012.35	72
80	1.490	.6710	.0102	.0152	98.068	65.802	36.848	2 424.65	80
84	1.520	.6577	.00961	.0146	104.074	68.453	38.576	2 640.67	84
90	1.567	.6383	.00883	.0138	113.311	72.331	41.145	2 976.08	90
96	1.614	.6195	.00814	.0131	122.829	76.095	43.685	3 324.19	96
100	1.647	.6073	.00773	.0127	129.334	78.543	45.361	3 562.80	100
104	1.680	.5953	.00735	.0124	135.970	80.942	47.025	3 806.29	104
120	1.819	.5496	.00610	.0111	163.880	90.074	53.551	4 823.52	120
240	3.310	.3021	.00216	.00716	462.041	139.581	96.113	13 415.56	240
360	6.023	.1660	.00100	.00600	1 004.5	166.792	128.324	21 403.32	360
480	10.957	.0913	.00050	.00550	1 991.5	181.748	151.795	27 588.37	480

3/4% **Compound Interest Factors** 3/4%

	Single Payment		Uniform Payment Series				Arithmetic Gradient		
	Compound Amount Factor	Present Worth Factor	Sinking Fund Factor	Capital Recovery Factor	Compound Amount Factor	Present Worth Factor	Gradient Uniform Series	Gradient Present Worth	
n	Find *F* Given *P* *F/P*	Find *P* Given *F* *P/F*	Find *A* Given *F* *A/F*	Find *A* Given *P* *A/P*	Find *F* Given *A* *F/A*	Find *P* Given *A* *P/A*	Find *A* Given *G* *A/G*	Find *P* Given *G* *P/G*	*n*
1	1.008	.9926	1.0000	1.0075	1.000	0.993	0	0	**1**
2	1.015	.9852	.4981	.5056	2.008	1.978	0.499	0.987	**2**
3	1.023	.9778	.3308	.3383	3.023	2.956	0.996	2.943	**3**
4	1.030	.9706	.2472	.2547	4.045	3.926	1.492	5.857	**4**
5	1.038	.9633	.1970	.2045	5.076	4.889	1.986	9.712	**5**
6	1.046	.9562	.1636	.1711	6.114	5.846	2.479	14.494	**6**
7	1.054	.9490	.1397	.1472	7.160	6.795	2.971	20.187	**7**
8	1.062	.9420	.1218	.1293	8.213	7.737	3.462	26.785	**8**
9	1.070	.9350	.1078	.1153	9.275	8.672	3.951	34.265	**9**
10	1.078	.9280	.0967	.1042	10.344	9.600	4.440	42.619	**10**
11	1.086	.9211	.0876	.0951	11.422	10.521	4.927	51.831	**11**
12	1.094	.9142	.0800	.0875	12.508	11.435	5.412	61.889	**12**
13	1.102	.9074	.0735	.0810	13.602	12.342	5.897	72.779	**13**
14	1.110	.9007	.0680	.0755	14.704	13.243	6.380	84.491	**14**
15	1.119	.8940	.0632	.0707	15.814	14.137	6.862	97.005	**15**
16	1.127	.8873	.0591	.0666	16.932	15.024	7.343	110.318	**16**
17	1.135	.8807	.0554	.0629	18.059	15.905	7.822	124.410	**17**
18	1.144	.8742	.0521	.0596	19.195	16.779	8.300	139.273	**18**
19	1.153	.8676	.0492	.0567	20.339	17.647	8.777	154.891	**19**
20	1.161	.8612	.0465	.0540	21.491	18.508	9.253	171.254	**20**
21	1.170	.8548	.0441	.0516	22.653	19.363	9.727	188.352	**21**
22	1.179	.8484	.0420	.0495	23.823	20.211	10.201	206.170	**22**
23	1.188	.8421	.0400	.0475	25.001	21.053	10.673	224.695	**23**
24	1.196	.8358	.0382	.0457	26.189	21.889	11.143	243.924	**24**
25	1.205	.8296	.0365	.0440	27.385	22.719	11.613	263.834	**25**
26	1.214	.8234	.0350	.0425	28.591	23.542	12.081	284.421	**26**
27	1.224	.8173	.0336	.0411	29.805	24.360	12.548	305.672	**27**
28	1.233	.8112	.0322	.0397	31.029	25.171	13.014	327.576	**28**
29	1.242	.8052	.0310	.0385	32.261	25.976	13.479	350.122	**29**
30	1.251	.7992	.0298	.0373	33.503	26.775	13.942	373.302	**30**
36	1.309	.7641	.0243	.0318	41.153	31.447	16.696	525.038	**36**
40	1.348	.7416	.0215	.0290	46.447	34.447	18.507	637.519	**40**
48	1.431	.6986	.0174	.0249	57.521	40.185	22.070	886.899	**48**
50	1.453	.6882	.0166	.0241	60.395	41.567	22.949	953.911	**50**
52	1.475	.6780	.0158	.0233	63.312	42.928	23.822	1 022.64	**52**
60	1.566	.6387	.0133	.0208	75.425	48.174	27.268	1 313.59	**60**
70	1.687	.5927	.0109	.0184	91.621	54.305	31.465	1 708.68	**70**
72	1.713	.5839	.0105	.0180	95.008	55.477	32.289	1 791.33	**72**
80	1.818	.5500	.00917	.0167	109.074	59.995	35.540	2 132.23	**80**
84	1.873	.5338	.00859	.0161	116.428	62.154	37.137	2 308.22	**84**
90	1.959	.5104	.00782	.0153	127.881	65.275	39.496	2 578.09	**90**
96	2.049	.4881	.00715	.0147	139.858	68.259	41.812	2 854.04	**96**
100	2.111	.4737	.00675	.0143	148.147	70.175	43.332	3 040.85	**100**
104	2.175	.4597	.00638	.0139	156.687	72.035	44.834	3 229.60	**104**
120	2.451	.4079	.00517	.0127	193.517	78.942	50.653	3 998.68	**120**
240	6.009	.1664	.00150	.00900	667.901	111.145	85.422	9 494.26	**240**
360	14.731	.0679	.00055	.00805	1 830.8	124.282	107.115	13 312.50	**360**
480	36.111	.0277	.00021	.00771	4 681.5	129.641	119.662	15 513.16	**480**

1% **Compound Interest Factors** 1%

	Single Payment		Uniform Payment Series				Arithmetic Gradient		
	Compound Amount Factor	Present Worth Factor	Sinking Fund Factor	Capital Recovery Factor	Compound Amount Factor	Present Worth Factor	Gradient Uniform Series	Gradient Present Worth	
	Find *F* Given *P*	Find *P* Given *F*	Find *A* Given *F*	Find *A* Given *P*	Find *F* Given *A*	Find *P* Given *A*	Find *A* Given *G*	Find *P* Given *G*	
n	*F/P*	*P/F*	*A/F*	*A/P*	*F/A*	*P/A*	*A/G*	*P/G*	***n***
1	1.010	.9901	1.0000	1.0100	1.000	0.990	0	0	**1**
2	1.020	.9803	.4975	.5075	2.010	1.970	0.498	0.980	**2**
3	1.030	.9706	.3300	.3400	3.030	2.941	0.993	2.921	**3**
4	1.041	.9610	.2463	.2563	4.060	3.902	1.488	5.804	**4**
5	1.051	.9515	.1960	.2060	5.101	4.853	1.980	9.610	**5**
6	1.062	.9420	.1625	.1725	6.152	5.795	2.471	14.320	**6**
7	1.072	.9327	.1386	.1486	7.214	6.728	2.960	19.917	**7**
8	1.083	.9235	.1207	.1307	8.286	7.652	3.448	26.381	**8**
9	1.094	.9143	.1067	.1167	9.369	8.566	3.934	33.695	**9**
10	1.105	.9053	.0956	.1056	10.462	9.471	4.418	41.843	**10**
11	1.116	.8963	.0865	.0965	11.567	10.368	4.900	50.806	**11**
12	1.127	.8874	.0788	.0888	12.682	11.255	5.381	60.568	**12**
13	1.138	.8787	.0724	.0824	13.809	12.134	5.861	71.112	**13**
14	1.149	.8700	.0669	.0769	14.947	13.004	6.338	82.422	**14**
15	1.161	.8613	.0621	.0721	16.097	13.865	6.814	94.481	**15**
16	1.173	.8528	.0579	.0679	17.258	14.718	7.289	107.273	**16**
17	1.184	.8444	.0543	.0643	18.430	15.562	7.761	120.783	**17**
18	1.196	.8360	.0510	.0610	19.615	16.398	8.232	134.995	**18**
19	1.208	.8277	.0481	.0581	20.811	17.226	8.702	149.895	**19**
20	1.220	.8195	.0454	.0554	22.019	18.046	9.169	165.465	**20**
21	1.232	.8114	.0430	.0530	23.239	18.857	9.635	181.694	**21**
22	1.245	.8034	.0409	.0509	24.472	19.660	10.100	198.565	**22**
23	1.257	.7954	.0389	.0489	25.716	20.456	10.563	216.065	**23**
24	1.270	.7876	.0371	.0471	26.973	21.243	11.024	234.179	**24**
25	1.282	.7798	.0354	.0454	28.243	22.023	11.483	252.892	**25**
26	1.295	.7720	.0339	.0439	29.526	22.795	11.941	272.195	**26**
27	1.308	.7644	.0324	.0424	30.821	23.560	12.397	292.069	**27**
28	1.321	.7568	.0311	.0411	32.129	24.316	12.852	312.504	**28**
29	1.335	.7493	.0299	.0399	33.450	25.066	13.304	333.486	**29**
30	1.348	.7419	.0287	.0387	34.785	25.808	13.756	355.001	**30**
36	1.431	.6989	.0232	.0332	43.077	30.107	16.428	494.620	**36**
40	1.489	.6717	.0205	.0305	48.886	32.835	18.178	596.854	**40**
48	1.612	.6203	.0163	.0263	61.223	37.974	21.598	820.144	**48**
50	1.645	.6080	.0155	.0255	64.463	39.196	22.436	879.417	**50**
52	1.678	.5961	.0148	.0248	67.769	40.394	23.269	939.916	**52**
60	1.817	.5504	.0122	.0222	81.670	44.955	26.533	1 192.80	**60**
70	2.007	.4983	.00993	.0199	100.676	50.168	30.470	1 528.64	**70**
72	2.047	.4885	.00955	.0196	104.710	51.150	31.239	1 597.86	**72**
80	2.217	.4511	.00822	.0182	121.671	54.888	34.249	1 879.87	**80**
84	2.307	.4335	.00765	.0177	130.672	56.648	35.717	2 023.31	**84**
90	2.449	.4084	.00690	.0169	144.863	59.161	37.872	2 240.56	**90**
96	2.599	.3847	.00625	.0163	159.927	61.528	39.973	2 459.42	**96**
100	2.705	.3697	.00587	.0159	170.481	63.029	41.343	2 605.77	**100**
104	2.815	.3553	.00551	.0155	181.464	64.471	42.688	2 752.17	**104**
120	3.300	.3030	.00435	.0143	230.039	69.701	47.835	3 334.11	**120**
240	10.893	.0918	.00101	.0110	989.254	90.819	75.739	6 878.59	**240**
360	35.950	.0278	.00029	.0103	3 495.0	97.218	89.699	8 720.43	**360**
480	118.648	.00843	.00008	.0101	11 764.8	99.157	95.920	9 511.15	**480**

1 1/4% **Compound Interest Factors** 1 1/4%

	Single Payment		Uniform Payment Series				Arithmetic Gradient		
	Compound Amount Factor	Present Worth Factor	Sinking Fund Factor	Capital Recovery Factor	Compound Amount Factor	Present Worth Factor	Gradient Uniform Series	Gradient Present Worth	
n	Find *F* Given *P* *F/P*	Find *P* Given *F* *P/F*	Find *A* Given *F* *A/F*	Find *A* Given *P* *A/P*	Find *F* Given *A* *F/A*	Find *P* Given *A* *P/A*	Find *A* Given *G* *A/G*	Find *P* Given *G* *P/G*	**n**
1	1.013	.9877	1.0000	1.0125	1.000	0.988	0	0	**1**
2	1.025	.9755	.4969	.5094	2.013	1.963	0.497	0.976	**2**
3	1.038	.9634	.3292	.3417	3.038	2.927	0.992	2.904	**3**
4	1.051	.9515	.2454	.2579	4.076	3.878	1.485	5.759	**4**
5	1.064	.9398	.1951	.2076	5.127	4.818	1.976	9.518	**5**
6	1.077	.9282	.1615	.1740	6.191	5.746	2.464	14.160	**6**
7	1.091	.9167	.1376	.1501	7.268	6.663	2.951	19.660	**7**
8	1.104	.9054	.1196	.1321	8.359	7.568	3.435	25.998	**8**
9	1.118	.8942	.1057	.1182	9.463	8.462	3.918	33.152	**9**
10	1.132	.8832	.0945	.1070	10.582	9.346	4.398	41.101	**10**
11	1.146	.8723	.0854	.0979	11.714	10.218	4.876	49.825	**11**
12	1.161	.8615	.0778	.0903	12.860	11.079	5.352	59.302	**12**
13	1.175	.8509	.0713	.0838	14.021	11.930	5.827	69.513	**13**
14	1.190	.8404	.0658	.0783	15.196	12.771	6.299	80.438	**14**
15	1.205	.8300	.0610	.0735	16.386	13.601	6.769	92.058	**15**
16	1.220	.8197	.0568	.0693	17.591	14.420	7.237	104.355	**16**
17	1.235	.8096	.0532	.0657	18.811	15.230	7.702	117.309	**17**
18	1.251	.7996	.0499	.0624	20.046	16.030	8.166	130.903	**18**
19	1.266	.7898	.0470	.0595	21.297	16.849	8.628	145.119	**19**
20	1.282	.7800	.0443	.0568	22.563	17.599	9.088	159.940	**20**
21	1.298	.7704	.0419	.0544	23.845	18.370	9.545	175.348	**21**
22	1.314	.7609	.0398	.0523	25.143	19.131	10.001	191.327	**22**
23	1.331	.7515	.0378	.0503	26.458	19.882	10.455	207.859	**23**
24	1.347	.7422	.0360	.0485	27.788	20.624	10.906	224.930	**24**
25	1.364	.7330	.0343	.0468	29.136	21.357	11.355	242.523	**25**
26	1.381	.7240	.0328	.0453	30.500	22.081	11.803	260.623	**26**
27	1.399	.7150	.0314	.0439	31.881	22.796	12.248	279.215	**27**
28	1.416	.7062	.0300	.0425	32.280	23.503	12.691	298.284	**28**
29	1.434	.6975	.0288	.0413	34.696	24.200	13.133	317.814	**29**
30	1.452	.6889	.0277	.0402	36.129	24.889	13.572	337.792	**30**
36	1.564	.6394	.0222	.0347	45.116	28.847	16.164	466.297	**36**
40	1.644	.6084	.0194	.0319	51.490	31.327	17.852	559.247	**40**
48	1.845	.5509	.0153	.0278	65.229	35.932	21.130	759.248	**48**
50	1.861	.5373	.0145	.0270	68.882	37.013	21.930	811.692	**50**
52	1.908	.5242	.0138	.0263	72.628	38.068	22.722	864.960	**52**
60	2.107	.4746	.0113	.0238	88.575	42.035	25.809	1 084.86	**60**
70	2.386	.4191	.00902	.0215	110.873	46.470	29.492	1 370.47	**70**
72	2.446	.4088	.00864	.0211	115.675	47.293	30.205	1 428.48	**72**
80	2.701	.3702	.00735	.0198	136.120	50.387	32.983	1 661.89	**80**
84	2.839	.3522	.00680	.0193	147.130	51.822	34.326	1 778.86	**84**
90	3.059	.3269	.00607	.0186	164.706	53.846	36.286	1 953.85	**90**
96	3.296	.3034	.00545	.0179	183.643	55.725	38.180	2 127.55	**96**
100	3.463	.2887	.00507	.0176	197.074	56.901	39.406	2 242.26	**100**
104	3.640	.2747	.00474	.0172	211.190	58.021	40.604	2 355.90	**104**
120	4.440	.2252	.00363	.0161	275.220	61.983	45.119	2 796.59	**120**
240	19.716	.0507	.00067	.0132	1 497.3	75.942	67.177	5 101.55	**240**
360	87.543	.0114	.00014	.0126	6 923.4	79.086	75.840	5 997.91	**360**
480	388.713	.00257	.00003	.0125	31 017.1	79.794	78.762	6 284.74	**480**

1 1/2% **Compound Interest Factors** 1 1/2%

	Single Payment		Uniform Payment Series				Arithmetic Gradient		
	Compound Amount Factor	Present Worth Factor	Sinking Fund Factor	Capital Recovery Factor	Compound Amount Factor	Present Worth Factor	Gradient Uniform Series	Gradient Present Worth	
n	Find *F* Given *P* *F/P*	Find *P* Given *F* *P/F*	Find *A* Given *F* *A/F*	Find *A* Given *P* *A/P*	Find *F* Given *A* *F/A*	Find *P* Given *A* *P/A*	Find *A* Given *G* *A/G*	Find *P* Given *G* *P/G*	**n**
1	1.015	.9852	1.0000	1.0150	1.000	0.985	0	0	**1**
2	1.030	.9707	.4963	.5113	2.015	1.956	0.496	0.970	**2**
3	1.046	.9563	.3284	.3434	3.045	2.912	0.990	2.883	**3**
4	1.061	.9422	.2444	.2594	4.091	3.854	1.481	5.709	**4**
5	1.077	.9283	.1941	.2091	5.152	4.783	1.970	9.422	**5**
6	1.093	.9145	.1605	.1755	6.230	5.697	2.456	13.994	**6**
7	1.110	.9010	.1366	.1516	7.323	6.598	2.940	19.400	**7**
8	1.126	.8877	.1186	.1336	8.433	7.486	3.422	25.614	**8**
9	1.143	.8746	.1046	.1196	9.559	8.360	3.901	32.610	**9**
10	1.161	.8617	.0934	.1084	10.703	9.222	4.377	40.365	**10**
11	1.178	.8489	.0843	.0993	11.863	10.071	4.851	48.855	**11**
12	1.196	.8364	.0767	.0917	13.041	10.907	5.322	58.054	**12**
13	1.214	.8240	.0702	.0852	14.237	11.731	5.791	67.943	**13**
14	1.232	.8118	.0647	.0797	15.450	12.543	6.258	78.496	**14**
15	1.250	.7999	.0599	.0749	16.682	13.343	6.722	89.694	**15**
16	1.269	.7880	.0558	.0708	17.932	14.131	7.184	101.514	**16**
17	1.288	.7764	.0521	.0671	19.201	14.908	7.643	113.937	**17**
18	1.307	.7649	.0488	.0638	20.489	15.673	8.100	126.940	**18**
19	1.327	.7536	.0459	.0609	21.797	16.426	8.554	140.505	**19**
20	1.347	.7425	.0432	.0582	23.124	17.169	9.005	154.611	**20**
21	1.367	.7315	.0409	.0559	24.470	17.900	9.455	169.241	**21**
22	1.388	.7207	.0387	.0537	25.837	18.621	9.902	184.375	**22**
23	1.408	.7100	.0367	.0517	27.225	19.331	10.346	199.996	**23**
24	1.430	.6995	.0349	.0499	28.633	20.030	10.788	216.085	**24**
25	1.451	.6892	.0333	.0483	30.063	20.720	11.227	232.626	**25**
26	1.473	.6790	.0317	.0467	31.514	21.399	11.664	249.601	**26**
27	1.495	.6690	.0303	.0453	32.987	22.068	12.099	266.995	**27**
28	1.517	.6591	.0290	.0440	34.481	22.727	12.531	284.790	**28**
29	1.540	.6494	.0278	.0428	35.999	23.376	12.961	302.972	**29**
30	1.563	.6398	.0266	.0416	37.539	24.016	13.388	321.525	**30**
36	1.709	.5851	.0212	.0362	47.276	27.661	15.901	439.823	**36**
40	1.814	.5513	.0184	.0334	54.268	29.916	17.528	524.349	**40**
48	2.043	.4894	.0144	.0294	69.565	34.042	20.666	703.537	**48**
50	2.105	.4750	.0136	.0286	73.682	35.000	21.428	749.955	**50**
52	2.169	.4611	.0128	.0278	77.925	35.929	22.179	796.868	**52**
60	2.443	.4093	.0104	.0254	96.214	39.380	25.093	988.157	**60**
70	2.835	.3527	.00817	.0232	122.363	43.155	28.529	1 231.15	**70**
72	2.921	.3423	.00781	.0228	128.076	43.845	29.189	1 279.78	**72**
80	3.291	.3039	.00655	.0215	152.710	46.407	31.742	1 473.06	**80**
84	3.493	.2863	.00602	.0210	166.172	47.579	32.967	1 568.50	**84**
90	3.819	.2619	.00532	.0203	187.929	49.210	34.740	1 709.53	**90**
96	4.176	.2395	.00472	.0197	211.719	50.702	36.438	1 847.46	**96**
100	4.432	.2256	.00437	.0194	228.802	51.625	37.529	1 937.43	**100**
104	4.704	.2126	.00405	.0190	246.932	52.494	38.589	2 025.69	**104**
120	5.969	.1675	.00302	.0180	331.286	55.498	42.518	2 359.69	**120**
240	35.632	.0281	.00043	.0154	2 308.8	64.796	59.737	3 870.68	**240**
360	212.700	.00470	.00007	.0151	14 113.3	66.353	64.966	4 310.71	**360**
480	1 269.7	.00079	.00001	.0150	84 577.8	66.614	66.288	4 415.74	**480**

1 3/4% **Compound Interest Factors** 1 3/4%

	Single Payment		Uniform Payment Series				Arithmetic Gradient		
	Compound Amount Factor	Present Worth Factor	Sinking Fund Factor	Capital Recovery Factor	Compound Amount Factor	Present Worth Factor	Gradient Uniform Series	Gradient Present Worth	
n	Find F Given P F/P	Find P Given F P/F	Find A Given F A/F	Find A Given P A/P	Find F Given A F/A	Find P Given A P/A	Find A Given G A/G	Find P Given G P/G	n
1	1.018	.9828	1.0000	1.0175	1.000	0.983	0	0	1
2	1.035	.9659	.4957	.5132	2.018	1.949	0.496	0.966	2
3	1.053	.9493	.3276	.3451	3.053	2.898	0.989	2.865	3
4	1.072	.9330	.2435	.2610	4.106	3.831	1.478	5.664	4
5	1.091	.9169	.1931	.2106	5.178	4.748	1.965	9.332	5
6	1.110	.9011	.1595	.1770	6.269	5.649	2.450	13.837	6
7	1.129	.8856	.1355	.1530	7.378	6.535	2.931	19.152	7
8	1.149	.8704	.1175	.1350	8.508	7.405	3.409	25.245	8
9	1.169	.8554	.1036	.1211	9.656	8.261	3.885	32.088	9
10	1.189	.8407	.0924	.1099	10.825	9.101	4.357	39.655	10
11	1.210	.8263	.0832	.1007	12.015`	9.928	4.827	47.918	11
12	1.231	.8121	.0756	.0931	13.225	10.740	5.294	56.851	12
13	1.253	.7981	.0692	.0867	14.457	11.538	5.758	66.428	13
14	1.275	.7844	.0637	.0812	15.710	12.322	6.219	76.625	14
15	1.297	.7709	.0589	.0764	16.985	13.093	6.677	87.417	15
16	1.320	.7576	.0547	.0722	18.282	13.851	7.132	98.782	16
17	1.343	.7446	.0510	.0685	19.602	14.595	7.584	110.695	17
18	1.367	.7318	.0477	.0652	20.945	15.327	8.034	123.136	18
19	1.390	.7192	.0448	.0623	22.311	16.046	8.481	136.081	19
20	1.415	.7068	.0422	.0597	23.702	16.753	8.924	149.511	20
21	1.440	.6947	.0398	.0573	25.116	17.448	9.365	163.405	21
22	1.465	.6827	.0377	.0552	26.556	18.130	9.804	177.742	22
23	1.490	.6710	.0357	.0532	28.021	18.801	10.239	192.503	23
24	1.516	.6594	.0339	.0514	29.511	19.461	10.671	207.671	24
25	1.543	.6481	.0322	.0497	31.028	20.109	11.101	223.225	25
26	1.570	.6369	.0307	.0482	32.571	20.746	11.528	239.149	26
27	1.597	.6260	.0293	.0468	34.141	21.372	11.952	255.425	27
28	1.625	.6152	.0280	.0455	35.738	21.987	12.373	272.036	28
29	1.654	.6046	.0268	.0443	37.363	22.592	12.791	288.967	29
30	1.683	.5942	.0256	.0431	39.017	23.186	13.206	306.200	30
36	1.867	.5355	.0202	.0377	49.566	26.543	15.640	415.130	36
40	2.002	.4996	.0175	.0350	57.234	28.594	17.207	492.017	40
48	2.300	.4349	.0135	.0310	74.263	32.294	20.209	652.612	48
50	2.381	.4200	.0127	.0302	78.903	33.141	20.932	693.708	50
52	2.465	.4057	.0119	.0294	83.706	33.960	21.644	735.039	52
60	2.832	.3531	.00955	.0271	104.676	36.964	24.389	901.503	60
70	3.368	.2969	.00739	.0249	135.331	40.178	27.586	1 108.34	70
72	3.487	.2868	.00704	.0245	142.127	40.757	28.195	1 149.12	72
80	4.006	.2496	.00582	.0233	171.795	42.880	30.533	1 309.25	80
84	4.294	.2329	.00531	.0228	188.246	43.836	31.644	1 387.16	84
90	4.765	.2098	.00465	.0221	215.166	45.152	33.241	1 500.88	90
96	5.288	.1891	.00408	.0216	245.039	46.337	34.756	1 610.48	96
100	5.668	.1764	.00375	.0212	266.753	47.062	35.721	1 681.09	100
104	6.075	.1646	.00345	.0209	290.028	47.737	36.652	1 749.68	104
120	8.019	.1247	.00249	.0200	401.099	50.017	40.047	2 003.03	120
240	64.308	.0156	.00028	.0178	3 617.6	56.254	53.352	3 001.27	240
360	515.702	.00194	.00003	.0175	29 411.5	57.032	56.443	3 219.08	360
480	4 135.5	.00024		.0175	236 259.0	57.129	57.027	3 257.88	480

2% **Compound Interest Factors** 2%

	Single Payment		Uniform Payment Series				Arithmetic Gradient		
n	Compound Amount Factor Find *F* Given *P* *F/P*	Present Worth Factor Find *P* Given *F* *P/F*	Sinking Fund Factor Find *A* Given *F* *A/F*	Capital Recovery Factor Find *A* Given *P* *A/P*	Compound Amount Factor Find *F* Given *A* *F/A*	Present Worth Factor Find *P* Given *A* *P/A*	Gradient Uniform Series Find *A* Given *G* *A/G*	Gradient Present Worth Find *P* Given *G* *P/G*	n
1	1.020	.9804	1.0000	1.0200	1.000	0.980	0	0	**1**
2	1.040	.9612	.4951	.5151	2.020	1.942	0.495	0.961	**2**
3	1.061	.9423	.3268	.3468	3.060	2.884	0.987	2.846	**3**
4	1.082	.9238	.2426	.2626	4.122	3.808	1.475	5.617	**4**
5	1.104	.9057	.1922	.2122	5.204	4.713	1.960	9.240	**5**
6	1.126	.8880	.1585	.1785	6.308	5.601	2.442	13.679	**6**
7	1.149	.8706	.1345	.1545	7.434	6.472	2.921	18.903	**7**
8	1.172	.8535	.1165	.1365	8.583	7.325	3.396	24.877	**8**
9	1.195	.8368	.1025	.1225	9.755	8.162	3.868	31.571	**9**
10	1.219	.8203	.0913	.1113	10.950	8.983	4.337	38.954	**10**
11	1.243	.8043	.0822	.1022	12.169	9.787	4.802	46.996	**11**
12	1.268	.7885	.0746	.0946	13.412	10.575	5.264	55.669	**12**
13	1.294	.7730	.0681	.0881	14.680	11.348	5.723	64.946	**13**
14	1.319	.7579	.0626	.0826	15.974	12.106	6.178	74.798	**14**
15	1.346	.7430	.0578	.0778	17.293	12.849	6.631	85.200	**15**
16	1.373	.7284	.0537	.0737	18.639	13.578	7.080	96.127	**16**
17	1.400	.7142	.0500	.0700	20.012	14.292	7.526	107.553	**17**
18	1.428	.7002	.0467	.0667	21.412	14.992	7.968	119.456	**18**
19	1.457	.6864	.0438	.0638	22.840	15.678	8.407	131.812	**19**
20	1.486	.6730	.0412	.0612	24.297	16.351	8.843	144.598	**20**
21	1.516	.6598	.0388	.0588	25.783	17.011	9.276	157.793	**21**
22	1.546	.6468	.0366	.0566	27.299	17.658	9.705	171.377	**22**
23	1.577	.6342	.0347	.0547	28.845	18.292	10.132	185.328	**23**
24	1.608	.6217	.0329	.0529	30.422	18.914	10.555	199.628	**24**
25	1.641	.6095	.0312	.0512	32.030	19.523	10.974	214.256	**25**
26	1.673	.5976	.0297	.0497	33.671	20.121	11.391	229.196	**26**
27	1.707	.5859	.0283	.0483	35.344	20.707	11.804	244.428	**27**
28	1.741	.5744	.0270	.0470	37.051	21.281	12.214	259.936	**28**
29	1.776	.5631	.0258	.0458	38.792	21.844	12.621	275.703	**29**
30	1.811	.5521	.0247	.0447	40.568	22.396	13.025	291.713	**30**
36	2.040	.4902	.0192	.0392	51.994	25.489	15.381	392.036	**36**
40	2.208	.4529	.0166	.0366	60.402	27.355	16.888	461.989	**40**
48	2.587	.3865	.0126	.0326	79.353	30.673	19.755	605.961	**48**
50	2.692	.3715	.0118	.0318	84.579	31.424	20.442	642.355	**50**
52	2.800	.3571	.0111	.0311	90.016	32.145	21.116	678.779	**52**
60	3.281	.3048	.00877	.0288	114.051	34.761	23.696	823.692	**60**
70	4.000	.2500	.00667	.0267	149.977	37.499	26.663	999.829	**70**
72	4.161	.2403	.00633	.0263	158.056	37.984	27.223	1 034.050	**72**
80	4.875	.2051	.00516	.0252	193.771	39.744	29.357	1 166.781	**80**
84	5.277	.1895	.00468	.0247	213.865	40.525	30.361	1 230.413	**84**
90	5.943	.1683	.00405	.0240	247.155	41.587	31.793	1 322.164	**90**
96	6.693	.1494	.00351	.0235	284.645	42.529	33.137	1 409.291	**96**
100	7.245	.1380	.00320	.0232	312.230	43.098	33.986	1 464.747	**100**
104	7.842	.1275	.00292	.0229	342.090	43.624	34.799	1 518.082	**104**
120	10.765	.0929	.00205	.0220	488.255	45.355	37.711	1 710.411	**120**
240	115.887	.00863	.00017	.0202	5 744.4	49.569	47.911	2 374.878	**240**
360	1 247.5	.00080	.00002	.0200	62 326.8	49.960	49.711	2 483.567	**360**
480	13 429.8	.00007		.0200	671 442.0	49.996	49.964	2 498.027	**480**

2 1/2% **Compound Interest Factors** 2 1/2%

	Single Payment		Uniform Payment Series				Arithmetic Gradient		
	Compound Amount Factor	Present Worth Factor	Sinking Fund Factor	Capital Recovery Factor	Compound Amount Factor	Present Worth Factor	Gradient Uniform Series	Gradient Present Worth	
	Find *F* Given *P*	Find *P* Given *F*	Find *A* Given *F*	Find *A* Given *P*	Find *F* Given *A*	Find *P* Given *A*	Find *A* Given *G*	Find *P* Given *G*	
n	*F/P*	*P/F*	*A/F*	*A/P*	*F/A*	*P/A*	*A/G*	*P/G*	*n*
1	1.025	.9756	1.0000	1.0250	1.000	0.976	0	0	1
2	1.051	.9518	.4938	.5188	2.025	1.927	0.494	0.952	2
3	1.077	.9286	.3251	.3501	3.076	2.856	0.984	2.809	3
4	1.104	.9060	.2408	.2658	4.153	3.762	1.469	5.527	4
5	1.131	.8839	.1902	.2152	5.256	4.646	1.951	9.062	5
6	1.160	.8623	.1566	.1816	6.388	5.508	2.428	13.374	6
7	1.189	.8413	.1325	.1575	7.547	6.349	2.901	18.421	7
8	1.218	.8207	.1145	.1395	8.736	7.170	3.370	24.166	8
9	1.249	.8007	.1005	.1255	9.955	7.971	3.835	30.572	9
10	1.280	.7812	.0893	.1143	11.203	8.752	4.296	37.603	10
11	1.312	.7621	.0801	.1051	12.483	9.514	4.753	45.224	11
12	1.345	.7436	.0725	.0975	13.796	10.258	5.206	53.403	12
13	1.379	.7254	.0660	.0910	15.140	10.983	5.655	62.108	13
14	1.413	.7077	.0605	.0855	16.519	11.691	6.100	71.309	14
15	1.448	.6905	.0558	.0808	17.932	12.381	6.540	80.975	15
16	1.485	.6736	.0516	.0766	19.380	13.055	6.977	91.080	16
17	1.522	.6572	.0479	.0729	20.865	13.712	7.409	101.595	17
18	1.560	.6412	.0447	.0697	22.386	14.353	7.838	112.495	18
19	1.599	.6255	.0418	.0668	23.946	14.979	8.262	123.754	19
20	1.639	.6103	.0391	.0641	25.545	15.589	8.682	135.349	20
21	1.680	.5954	.0368	.0618	27.183	16.185	9.099	147.257	21
22	1.722	.5809	.0346	.0596	28.863	16.765	9.511	159.455	22
23	1.765	.5667	.0327	.0577	30.584	17.332	9.919	171.922	23
24	1.809	.5529	.0309	.0559	32.349	17.885	10.324	184.638	24
25	1.854	.5394	.0293	.0543	34.158	18.424	10.724	197.584	25
26	1.900	.5262	.0278	.0528	36.012	18.951	11.120	210.740	26
27	1.948	.5134	.0264	.0514	37.912	19.464	11.513	224.088	27
28	1.996	.5009	.0251	.0501	39.860	19.965	11.901	237.612	28
29	2.046	.4887	.0239	.0489	41.856	20.454	12.286	251.294	29
30	2.098	.4767	.0228	.0478	43.903	20.930	12.667	265.120	30
31	2.150	.4651	.0217	.0467	46.000	21.395	13.044	279.073	31
32	2.204	.4538	.0208	.0458	48.150	24.849	13.417	293.140	32
33	2.259	.4427	.0199	.0449	50.354	22.292	13.786	307.306	33
34	2.315	.4319	.0190	.0440	52.613	22.724	14.151	321.559	34
35	2.373	.4214	.0182	.0432	54.928	23.145	14.512	335.886	35
40	2.685	.3724	.0148	.0398	67.402	25.103	16.262	408.221	40
45	3.038	.3292	.0123	.0373	81.516	26.833	17.918	480.806	45
50	3.437	.2909	.0103	.0353	97.484	28.362	19.484	552.607	50
55	3.889	.2572	.00865	.0337	115.551	29.714	20.961	622.827	55
60	4.400	.2273	.00735	.0324	135.991	30.909	22.352	690.865	60
65	4.978	.2009	.00628	.0313	159.118	31.965	23.660	756.280	65
70	5.632	.1776	.00540	.0304	185.284	32.898	24.888	818.763	70
75	6.372	.1569	.00465	.0297	214.888	33.723	26.039	878.114	75
80	7.210	.1387	.00403	.0290	248.382	34.452	27.117	934.217	80
85	8.157	.1226	.00349	.0285	286.278	35.096	28.123	987.026	85
90	9.229	.1084	.00304	.0280	329.154	35.666	29.063	1 036.54	90
95	10.442	.0958	.00265	.0276	377.663	36.169	29.938	1 082.83	95
100	11.814	.0846	.00231	.0273	432.548	36.614	30.752	1 125.97	100

3% **Compound Interest Factors** 3%

	Single Payment		Uniform Payment Series				Arithmetic Gradient		
	Compound Amount Factor	Present Worth Factor	Sinking Fund Factor	Capital Recovery Factor	Compound Amount Factor	Present Worth Factor	Gradient Uniform Series	Gradient Present Worth	
	Find *F* Given *P*	Find *P* Given *F*	Find *A* Given *F*	Find *A* Given *P*	Find *F* Given *A*	Find *P* Given *A*	Find *A* Given *G*	Find *P* Given *G*	
n	*F/P*	*P/F*	*A/F*	*A/P*	*F/A*	*P/A*	*A/G*	*P/G*	*n*
1	1.030	.9709	1.0000	1.0300	1.000	0.971	0	0	**1**
2	1.061	.9426	.4926	.5226	2.030	1.913	0.493	0.943	**2**
3	1.093	.9151	.3235	.3535	3.091	2.829	0.980	2.773	**3**
4	1.126	.8885	.2390	.2690	4.184	3.717	1.463	5.438	**4**
5	1.159	.8626	.1884	.2184	5.309	4.580	1.941	8.889	**5**
6	1.194	.8375	.1546	.1846	6.468	5.417	2.414	13.076	**6**
7	1.230	.8131	.1305	.1605	7.662	6.230	2.882	17.955	**7**
8	1.267	.7894	.1125	.1425	8.892	7.020	3.345	23.481	**8**
9	1.305	.7664	.0984	.1284	10.159	7.786	3.803	29.612	**9**
10	1.344	.7441	.0872	.1172	11.464	8.530	4.256	36.309	**10**
11	1.384	.7224	.0781	.1081	12.808	9.253	4.705	43.533	**11**
12	1.426	.7014	.0705	.1005	14.192	9.954	5.148	51.248	**12**
13	1.469	.6810	.0640	.0940	15.618	10.635	5.587	59.419	**13**
14	1.513	.6611	.0585	.0885	17.086	11.296	6.021	68.014	**14**
15	1.558	.6419	.0538	.0838	18.599	11.938	6.450	77.000	**15**
16	1.605	.6232	.0496	.0796	20.157	12.561	6.874	86.348	**16**
17	1.653	.6050	.0460	.0760	21.762	13.166	7.294	96.028	**17**
18	1.702	.5874	.0427	.0727	23.414	13.754	7.708	106.014	**18**
19	1.754	.5703	.0398	.0698	25.117	14.324	8.118	116.279	**19**
20	1.806	.5537	.0372	.0672	26.870	14.877	8.523	126.799	**20**
21	1.860	.5375	.0349	.0649	28.676	15.415	8.923	137.549	**21**
22	1.916	.5219	.0327	.0627	30.537	15.937	9.319	148.509	**22**
23	1.974	.5067	.0308	.0608	32.453	16.444	9.709	159.656	**23**
24	2.033	.4919	.0290	.0590	34.426	16.936	10.095	170.971	**24**
25	2.094	.4776	.0274	.0574	36.459	17.413	10.477	182.433	**25**
26	2.157	.4637	.0259	.0559	38.553	17.877	10.853	194.026	**26**
27	2.221	.4502	.0246	.0546	40.710	18.327	11.226	205.731	**27**
28	2.288	.4371	.0233	.0533	42.931	18.764	11.593	217.532	**28**
29	2.357	.4243	.0221	.0521	45.219	19.188	11.956	229.413	**29**
30	2.427	.4120	.0210	.0510	47.575	19.600	12.314	241.361	**30**
31	2.500	.4000	.0200	.0500	50.003	20.000	12.668	253.361	**31**
32	2.575	.3883	.0190	.0490	52.503	20.389	13.017	265.399	**32**
33	2.652	.3770	.0182	.0482	55.078	20.766	13.362	277.464	**33**
34	2.732	.3660	.0173	.0473	57.730	21.132	13.702	289.544	**34**
35	2.814	.3554	.0165	.0465	60.462	21.487	14.037	301.627	**35**
40	3.262	.3066	.0133	.0433	75.401	23.115	15.650	361.750	**40**
45	3.782	.2644	.0108	.0408	92.720	24.519	17.156	420.632	**45**
50	4.384	.2281	.00887	.0389	112.797	25.730	18.558	477.480	**50**
55	5.082	.1968	.00735	.0373	136.072	26.774	19.860	531.741	**55**
60	5.892	.1697	.00613	.0361	163.053	27.676	21.067	583.052	**60**
65	6.830	.1464	.00515	.0351	194.333	28.453	22.184	631.201	**65**
70	7.918	.1263	.00434	.0343	230.594	29.123	23.215	676.087	**70**
75	9.179	.1089	.00367	.0337	272.631	29.702	24.163	717.698	**75**
80	10.641	.0940	.00311	.0331	321.363	30.201	25.035	756.086	**80**
85	12.336	.0811	.00265	.0326	377.857	30.631	25.835	791.353	**85**
90	14.300	.0699	.00226	.0323	443.349	31.002	26.567	823.630	**90**
95	16.578	.0603	.00193	.0319	519.272	31.323	27.235	853.074	**95**
100	19.219	.0520	.00165	.0316	607.287	31.599	27.844	879.854	**100**

3 1/2% **Compound Interest Factors** 3 1/2%

	Single Payment		Uniform Payment Series				Arithmetic Gradient		
	Compound Amount Factor	Present Worth Factor	Sinking Fund Factor	Capital Recovery Factor	Compound Amount Factor	Present Worth Factor	Gradient Uniform Series	Gradient Present Worth	
n	Find *F* Given *P* *F/P*	Find *P* Given *F* *P/F*	Find *A* Given *F* *A/F*	Find *A* Given *P* *A/P*	Find *F* Given *A* *F/A*	Find *P* Given *A* *P/A*	Find *A* Given *G* *A/G*	Find *P* Given *G* *P/G*	*n*
1	1.035	.9662	1.0000	1.0350	1.000	0.966	0	0	1
2	1.071	.9335	.4914	.5264	2.035	1.900	0.491	0.933	2
3	1.109	.9019	.3219	.3569	3.106	2.802	0.977	2.737	3
4	1.148	.8714	.2373	.2723	4.215	3.673	1.457	5.352	4
5	1.188	.8420	.1865	.2215	5.362	4.515	1.931	8.719	5
6	1.229	.8135	.1527	.1877	6.550	5.329	2.400	12.787	6
7	1.272	.7860	.1285	.1635	7.779	6.115	2.862	17.503	7
8	1.317	.7594	.1105	.1455	9.052	6.874	3.320	22.819	8
9	1.363	.7337	.0964	.1314	10.368	7.608	3.771	28.688	9
10	1.411	.7089	.0852	.1202	11.731	8.317	4.217	35.069	10
11	1.460	.6849	.0761	.1111	13.142	9.002	4.657	41.918	11
12	1.511	.6618	.0685	.1035	14.602	9.663	5.091	49.198	12
13	1.564	.6394	.0621	.0971	16.113	10.303	5.520	56.871	13
14	1.619	.6178	.0566	.0916	17.677	10.921	5.943	64.902	14
15	1.675	.5969	.0518	.0868	19.296	11.517	6.361	73.258	15
16	1.734	.5767	.0477	.0827	20.971	12.094	6.773	81.909	16
17	1.795	.5572	.0440	.0790	22.705	12.651	7.179	90.824	17
18	1.857	.5384	.0408	.0758	24.500	13.190	7.580	99.976	18
19	1.922	.5202	.0379	.0729	26.357	13.710	7.975	109.339	19
20	1.990	.5026	.0354	.0704	28.280	14.212	8.365	118.888	20
21	2.059	.4856	.0330	.0680	30.269	14.698	8.749	128.599	21
22	2.132	.4692	.0309	.0659	32.329	15.167	9.128	138.451	22
23	2.206	.4533	.0290	.0640	34.460	15.620	9.502	148.423	23
24	2.283	.4380	.0273	.0623	36.666	16.058	9.870	158.496	24
25	2.363	.4231	.0257	.0607	38.950	16.482	10.233	168.652	25
26	2.446	.4088	.0242	.0592	41.313	16.890	10.590	178.873	26
27	2.532	.3950	.0229	.0579	43.759	17.285	10.942	189.143	27
28	2.620	.3817	.0216	.0566	46.291	17.667	11.289	199.448	28
29	2.712	.3687	.0204	.0554	48.911	18.036	11.631	209.773	29
30	2.807	.3563	.0194	.0544	51.623	18.392	11.967	220.105	30
31	2.905	.3442	.0184	.0534	54.429	18.736	12.299	230.432	31
32	3.007	.3326	.0174	.0524	57.334	19.069	12.625	240.742	32
33	3.112	.3213	.0166	.0516	60.341	19.390	12.946	251.025	33
34	3.221	.3105	.0158	.0508	63.453	19.701	13.262	261.271	34
35	3.334	.3000	.0150	.0500	66.674	20.001	13.573	271.470	35
40	3.959	.2526	.0118	.0468	84.550	21.355	15.055	321.490	40
45	4.702	.2127	.00945	.0445	105.781	22.495	16.417	369.307	45
50	5.585	.1791	.00763	.0426	130.998	23.456	17.666	414.369	50
55	6.633	.1508	.00621	.0412	160.946	24.264	18.808	456.352	55
60	7.878	.1269	.00509	.0401	196.516	24.945	19.848	495.104	60
65	9.357	.1069	.00419	.0392	238.762	25.518	20.793	530.598	65
70	11.113	.0900	.00346	.0385	288.937	26.000	21.650	562.895	70
75	13.199	.0758	.00287	.0379	348.529	26.407	22.423	592.121	75
80	15.676	.0638	.00238	.0374	419.305	26.749	23.120	618.438	80
85	18.618	.0537	.00199	.0370	503.365	27.037	23.747	642.036	85
90	22.112	.0452	.00166	.0367	603.202	27.279	24.308	663.118	90
95	26.262	.0381	.00139	.0364	721.778	27.483	24.811	681.890	95
100	31.191	.0321	.00116	.0362	862.608	27.655	25.259	698.554	100

4% **Compound Interest Factors** 4%

	Single Payment		Uniform Payment Series				Arithmetic Gradient		
	Compound Amount Factor	Present Worth Factor	Sinking Fund Factor	Capital Recovery Factor	Compound Amount Factor	Present Worth Factor	Gradient Uniform Series	Gradient Present Worth	
n	**Find *F* Given *P* *F/P***	**Find *P* Given *F* *P/F***	**Find *A* Given *F* *A/F***	**Find *A* Given *P* *A/P***	**Find *F* Given *A* *F/A***	**Find *P* Given *A* *P/A***	**Find *A* Given *G* *A/G***	**Find *P* Given *G* *P/G***	***n***
1	1.040	.9615	1.0000	1.0400	1.000	0.962	0	0	**1**
2	1.082	.9246	.4902	.5302	2.040	1.886	0.490	0.925	**2**
3	1.125	.8890	.3203	.3603	3.122	2.775	0.974	2.702	**3**
4	1.170	.8548	.2355	.2755	4.246	3.630	1.451	5.267	**4**
5	1.217	.8219	.1846	.2246	5.416	4.452	1.922	8.555	**5**
6	1.265	.7903	.1508	.1908	6.633	5.242	2.386	12.506	**6**
7	1.316	.7599	.1266	.1666	7.898	6.002	2.843	17.066	**7**
8	1.369	.7307	.1085	.1485	9.214	6.733	3.294	22.180	**8**
9	1.423	.7026	.0945	.1345	10.583	7.435	3.739	27.801	**9**
10	1.480	.6756	.0833	.1233	12.006	8.111	4.177	33.881	**10**
11	1.539	.6496	.0741	.1141	13.486	8.760	4.609	40.377	**11**
12	1.601	.6246	.0666	.1066	15.026	9.385	5.034	47.248	**12**
13	1.665	.6006	.0601	.1001	16.627	9.986	5.453	54.454	**13**
14	1.732	.5775	.0547	.0947	18.292	10.563	5.866	61.962	**14**
15	1.801	.5553	.0499	.0899	20.024	11.118	6.272	69.735	**15**
16	1.873	.5339	.0458	.0858	21.825	11.652	6.672	77.744	**16**
17	1.948	.5134	.0422	.0822	23.697	12.166	7.066	85.958	**17**
18	2.026	.4936	.0390	.0790	25.645	12.659	7.453	94.350	**18**
19	2.107	.4746	.0361	.0761	27.671	13.134	7.834	102.893	**19**
20	2.191	.4564	.0336	.0736	29.778	13.590	8.209	111.564	**20**
21	2.279	.4388	.0313	.0713	31.969	14.029	8.578	120.341	**21**
22	2.370	.4220	.0292	.0692	34.248	14.451	8.941	129.202	**22**
23	2.465	.4057	.0273	.0673	36.618	14.857	9.297	138.128	**23**
24	2.563	.3901	.0256	.0656	39.083	15.247	9.648	147.101	**24**
25	2.666	.3751	.0240	.0640	41.646	15.622	9.993	156.104	**25**
26	2.772	.3607	.0226	.0626	44.312	15.983	10.331	165.121	**26**
27	2.883	.3468	.0212	.0612	47.084	16.330	10.664	174.138	**27**
28	2.999	.3335	.0200	.0600	49.968	16.663	10.991	183.142	**28**
29	3.119	.3207	.0189	.0589	52.966	16.984	11.312	192.120	**29**
30	3.243	.3083	.0178	.0578	56.085	17.292	11.627	201.062	**30**
31	3.373	.2965	.0169	.0569	59.328	17.588	11.937	209.955	**31**
32	3.508	.2851	.0159	.0559	62.701	17.874	12.241	218.792	**32**
33	3.648	.2741	.0151	.0551	66.209	18.148	12.540	227.563	**33**
34	3.794	.2636	.0143	.0543	69.858	18.411	12.832	236.260	**34**
35	3.946	.2534	.0136	.0536	73.652	18.665	13.120	244.876	**35**
40	4.801	.2083	.0105	.0505	95.025	19.793	14.476	286.530	**40**
45	5.841	.1712	.00826	.0483	121.029	20.720	15.705	325.402	**45**
50	7.107	.1407	.00655	.0466	152.667	21.482	16.812	361.163	**50**
55	8.646	.1157	.00523	.0452	191.159	22.109	17.807	393.689	**55**
60	10.520	.0951	.00420	.0442	237.990	22.623	18.697	422.996	**60**
65	12.799	.0781	.00339	.0434	294.968	23.047	19.491	449.201	**65**
70	15.572	.0642	.00275	.0427	364.290	23.395	20.196	472.479	**70**
75	18.945	.0528	.00223	.0422	448.630	23.680	20.821	493.041	**75**
80	23.050	.0434	.00181	.0418	551.244	23.915	21.372	511.116	**80**
85	28.044	.0357	.00148	.0415	676.089	24.109	21.857	526.938	**85**
90	34.119	.0293	.00121	.0412	827.981	24.267	22.283	540.737	**90**
95	41.511	.0241	.00099	.0410	1 012.8	24.398	22.655	552.730	**95**
100	50.505	.0198	.00081	.0408	1 237.6	24.505	22.980	563.125	**100**

4 1/2% **Compound Interest Factors** 4 1/2%

	Single Payment		Uniform Payment Series				Arithmetic Gradient		
	Compound Amount Factor	Present Worth Factor	Sinking Fund Factor	Capital Recovery Factor	Compound Amount Factor	Present Worth Factor	Gradient Uniform Series	Gradient Present Worth	
	Find F Given P	Find P Given F	Find A Given F	Find A Given P	Find F Given A	Find P Given A	Find A Given G	Find P Given G	
n	F/P	P/F	A/F	A/P	F/A	P/A	A/G	P/G	n
1	1.045	.9569	1.0000	1.0450	1.000	0.957	0	0	1
2	1.092	.9157	.4890	.5340	2.045	1.873	0.489	0.916	2
3	1.141	.8763	.3188	.3638	3.137	2.749	0.971	2.668	3
4	1.193	.8386	.2337	.2787	4.278	3.588	1.445	5.184	4
5	1.246	.8025	.1828	.2278	5.471	4.390	1.912	8.394	5
6	1.302	.7679	.1489	.1939	6.717	5.158	2.372	12.233	6
7	1.361	.7348	.1247	.1697	8.019	5.893	2.824	16.642	7
8	1.422	.7032	.1066	.1516	9.380	6.596	3.269	21.564	8
9	1.486	.6729	.0926	.1376	10.802	7.269	3.707	26.948	9
10	1.553	.6439	.0814	.1264	12.288	7.913	4.138	32.743	10
11	1.623	.6162	.0722	.1172	13.841	8.529	4.562	38.905	11
12	1.696	.5897	.0647	.1097	15.464	9.119	4.978	45.391	12
13	1.772	.5643	.0583	.1033	17.160	9.683	5.387	52.163	13
14	1.852	.5400	.0528	.0978	18.932	10.223	5.789	59.182	14
15	1.935	.5167	.0481	.0931	20.784	10.740	6.184	66.416	15
16	2.022	.4945	.0440	.0890	22.719	11.234	6.572	73.833	16
17	2.113	.4732	.0404	.0854	24.742	11.707	6.953	81.404	17
18	2.208	.4528	.0372	.0822	26.855	12.160	7.327	89.102	18
19	2.308	.4333	.0344	.0794	29.064	12.593	7.695	96.901	19
20	2.412	.4146	.0319	.0769	31.371	13.008	8.055	104.779	20
21	2.520	.3968	.0296	.0746	33.783	13.405	8.409	112.715	21
22	2.634	.3797	.0275	.0725	36.303	13.784	8.755	120.689	22
23	2.752	.3634	.0257	.0707	38.937	14.148	9.096	128.682	23
24	2.876	.3477	.0240	.0690	41.689	14.495	9.429	136.680	24
25	3.005	.3327	.0224	.0674	44.565	14.828	9.756	144.665	25
26	3.141	.3184	.0210	.0660	47.571	15.147	10.077	152.625	26
27	3.282	.3047	.0197	.0647	50.711	15.451	10.391	160.547	27
28	3.430	.2916	.0185	.0635	53.993	15.743	10.698	168.420	28
29	3.584	.2790	.0174	.0624	57.423	16.022	10.999	176.232	29
30	3.745	.2670	.0164	.0614	61.007	16.289	11.295	183.975	30
31	3.914	.2555	.0154	.0604	64.752	16.544	11.583	191.640	31
32	4.090	.2445	.0146	.0596	68.666	16.789	11.866	199.220	32
33	4.274	.2340	.0137	.0587	72.756	17.023	12.143	206.707	33
34	4.466	.2239	.0130	.0580	77.030	17.247	12.414	214.095	34
35	4.667	.2143	.0123	.0573	81.497	17.461	12.679	221.380	35
40	5.816	.1719	.00934	.0543	107.030	18.402	13.917	256.098	40
45	7.248	.1380	.00720	.0522	138.850	19.156	15.020	287.732	45
50	9.033	.1107	.00560	.0506	178.503	19.762	15.998	316.145	50
55	11.256	.0888	.00439	.0494	227.918	20.248	16.860	341.375	55
60	14.027	.0713	.00345	.0485	289.497	20.638	17.617	363.571	60
65	17.481	.0572	.00273	.0477	366.237	20.951	18.278	382.946	65
70	21.784	.0459	.00217	.0472	461.869	21.202	18.854	399.750	70
75	27.147	.0368	.00172	.0467	581.043	21.404	19.354	414.242	75
80	33.830	.0296	.00137	.0464	729.556	21.565	19.785	426.680	80
85	42.158	.0237	.00109	.0461	914.630	21.695	20.157	437.309	85
90	52.537	.0190	.00087	.0459	1 145.3	21.799	20.476	446.359	90
95	65.471	.0153	.00070	.0457	1 432.7	21.883	20.749	454.039	95
100	81.588	.0123	.00056	.0456	1 790.9	21.950	20.981	460.537	100

5% **Compound Interest Factors** 5%

	Single Payment		Uniform Payment Series				Arithmetic Gradient		
	Compound Amount Factor	Present Worth Factor	Sinking Fund Factor	Capital Recovery Factor	Compound Amount Factor	Present Worth Factor	Gradient Uniform Series	Gradient Present Worth	
n	Find *F* Given *P* *F/P*	Find *P* Given *F* *P/F*	Find *A* Given *F* *A/F*	Find *A* Given *P* *A/P*	Find *F* Given *A* *F/A*	Find *P* Given *A* *P/A*	Find *A* Given *G* *A/G*	Find *P* Given *G* *P/G*	*n*
1	1.050	.9524	1.0000	1.0500	1.000	0.952	0	0	**1**
2	1.102	.9070	.4878	.5378	2.050	1.859	0.488	0.907	**2**
3	1.158	.8638	.3172	.3672	3.152	2.723	0.967	2.635	**3**
4	1.216	.8227	.2320	.2820	4.310	3.546	1.439	5.103	**4**
5	1.276	.7835	.1810	.2310	5.526	4.329	1.902	8.237	**5**
6	1.340	.7462	.1470	.1970	6.802	5.076	2.358	11.968	**6**
7	1.407	.7107	.1228	.1728	8.142	5.786	2.805	16.232	**7**
8	1.477	.6768	.1047	.1547	9.549	6.463	3.244	20.970	**8**
9	1.551	.6446	.0907	.1407	11.027	7.108	3.676	26.127	**9**
10	1.629	.6139	.0795	.1295	12.578	7.722	4.099	31.652	**10**
11	1.710	.5847	.0704	.1204	14.207	8.306	4.514	37.499	**11**
12	1.796	.5568	.0628	.1128	15.917	8.863	4.922	43.624	**12**
13	1.886	.5303	.0565	.1065	17.713	9.394	5.321	49.988	**13**
14	1.980	.5051	.0510	.1010	19.599	9.899	5.713	56.553	**14**
15	2.079	.4810	.0463	.0963	21.579	10.380	6.097	63.288	**15**
16	2.183	.4581	.0423	.0923	23.657	10.838	6.474	70.159	**16**
17	2.292	.4363	.0387	.0887	25.840	11.274	6.842	77.140	**17**
18	2.407	.4155	.0355	.0855	28.132	11.690	7.203	84.204	**18**
19	2.527	.3957	.0327	.0827	30.539	12.085	7.557	91.327	**19**
20	2.653	.3769	.0302	.0802	33.066	12.462	7.903	98.488	**20**
21	2.786	.3589	.0280	.0780	35.719	12.821	8.242	105.667	**21**
22	2.925	.3419	.0260	.0760	38.505	13.163	8.573	112.846	**22**
23	3.072	.3256	.0241	.0741	41.430	13.489	8.897	120.008	**23**
24	3.225	.3101	.0225	.0725	44.502	13.799	9.214	127.140	**24**
25	3.386	.2953	.0210	.0710	47.727	14.094	9.524	134.227	**25**
26	3.556	.2812	.0196	.0696	51.113	14.375	9.827	141.258	**26**
27	3.733	.2678	.0183	.0683	54.669	14.643	10.122	148.222	**27**
28	3.920	.2551	.0171	.0671	58.402	14.898	10.411	155.110	**28**
29	4.116	.2429	.0160	.0660	62.323	15.141	10.694	161.912	**29**
30	4.322	.2314	.0151	.0651	66.439	15.372	10.969	168.622	**30**
31	4.538	.2204	.0141	.0641	70.761	15.593	11.238	175.233	**31**
32	4.765	.2099	.0133	.0633	75.299	15.803	11.501	181.739	**32**
33	5.003	.1999	.0125	.0625	80.063	16.003	11.757	188.135	**33**
34	5.253	.1904	.0118	.0618	85.067	16.193	12.006	194.416	**34**
35	5.516	.1813	.0111	.0611	90.320	16.374	12.250	200.580	**35**
40	7.040	.1420	.00828	.0583	120.799	17.159	13.377	229.545	**40**
45	8.985	.1113	.00626	.0563	159.699	17.774	14.364	255.314	**45**
50	11.467	.0872	.00478	.0548	209.347	18.256	15.223	277.914	**50**
55	14.636	.0683	.00367	.0537	272.711	18.633	15.966	297.510	**55**
60	18.679	.0535	.00283	.0528	353.582	18.929	16.606	314.343	**60**
65	23.840	.0419	.00219	.0522	456.795	19.161	17.154	328.691	**65**
70	30.426	.0329	.00170	.0517	588.525	19.343	17.621	340.841	**70**
75	38.832	.0258	.00132	.0513	756.649	19.485	18.018	351.072	**75**
80	49.561	.0202	.00103	.0510	971.222	19.596	18.353	359.646	**80**
85	63.254	.0158	.00080	.0508	1 245.1	19.684	18.635	366.800	**85**
90	80.730	.0124	.00063	.0506	1 594.6	19.752	18.871	372.749	**90**
95	103.034	.00971	.00049	.0505	2 040.7	19.806	19.069	377.677	**95**
100	131.500	.00760	.00038	.0504	2 610.0	19.848	19.234	381.749	**100**

6% **Compound Interest Factors** 6%

	Single Payment		Uniform Payment Series				Arithmetic Gradient		
	Compound Amount Factor	Present Worth Factor	Sinking Fund Factor	Capital Recovery Factor	Compound Amount Factor	Present Worth Factor	Gradient Uniform Series	Gradient Present Worth	
n	Find *F* Given *P* *F/P*	Find *P* Given *F* *P/F*	Find *A* Given *F* *A/F*	Find *A* Given *P* *A/P*	Find *F* Given *A* *F/A*	Find *P* Given *A* *P/A*	Find *A* Given *G* *A/G*	Find *P* Given *G* *P/G*	*n*
1	1.060	.9434	1.0000	1.0600	1.000	0.943	0	0	**1**
2	1.124	.8900	.4854	.5454	2.060	1.833	0.485	0.890	**2**
3	1.191	.8396	.3141	.3741	3.184	2.673	0.961	2.569	**3**
4	1.262	.7921	.2286	.2886	4.375	3.465	1.427	4.945	**4**
5	1.338	.7473	.1774	.2374	5.637	4.212	1.884	7.934	**5**
6	1.419	.7050	.1434	.2034	6.975	4.917	2.330	11.459	**6**
7	1.504	.6651	.1191	.1791	8.394	5.582	2.768	15.450	**7**
8	1.594	.6274	.1010	.1610	9.897	6.210	3.195	19.841	**8**
9	1.689	.5919	.0870	.1470	11.491	6.802	3.613	24.577	**9**
10	1.791	.5584	.0759	.1359	13.181	7.360	4.022	29.602	**10**
11	1.898	.5268	.0668	.1268	14.972	7.887	4.421	34.870	**11**
12	2.012	.4970	.0593	.1193	16.870	8.384	4.811	40.337	**12**
13	2.133	.4688	.0530	.1130	18.882	8.853	5.192	45.963	**13**
14	2.261	.4423	.0476	.1076	21.015	9.295	5.564	51.713	**14**
15	2.397	.4173	.0430	.1030	23.276	9.712	5.926	57.554	**15**
16	2.540	.3936	.0390	.0990	25.672	10.106	6.279	63.459	**16**
17	2.693	.3714	.0354	.0954	28.213	10.477	6.624	69.401	**17**
18	2.854	.3503	.0324	.0924	30.906	10.828	6.960	75.357	**18**
19	3.026	.3305	.0296	.0896	33.760	11.158	7.287	81.306	**19**
20	3.207	.3118	.0272	.0872	36.786	11.470	7.605	87.230	**20**
21	3.400	.2942	.0250	.0850	39.993	11.764	7.915	93.113	**21**
22	3.604	.2775	.0230	.0830	43.392	12.042	8.217	98.941	**22**
23	3.820	.2618	.0213	.0813	46.996	12.303	8.510	104.700	**23**
24	4.049	.2470	.0197	.0797	50.815	12.550	8.795	110.381	**24**
25	4.292	.2330	.0182	.0782	54.864	12.783	9.072	115.973	**25**
26	4.549	.2198	.0169	.0769	59.156	13.003	9.341	121.468	**26**
27	4.822	.2074	.0157	.0757	63.706	13.211	9.603	126.860	**27**
28	5.112	.1956	.0146	.0746	68.528	13.406	9.857	132.142	**28**
29	5.418	.1846	.0136	.0736	73.640	13.591	10.103	137.309	**29**
30	5.743	.1741	.0126	.0726	79.058	13.765	10.342	142.359	**30**
31	6.088	.1643	.0118	.0718	84.801	13.929	10.574	147.286	**31**
32	6.453	.1550	.0110	.0710	90.890	14.084	10.799	152.090	**32**
33	6.841	.1462	.0103	.0703	97.343	14.230	11.017	156.768	**33**
34	7.251	.1379	.00960	.0696	104.184	14.368	11.228	161.319	**34**
35	7.686	.1301	.00897	.0690	111.435	14.498	11.432	165.743	**35**
40	10.286	.0972	.00646	.0665	154.762	15.046	12.359	185.957	**40**
45	13.765	.0727	.00470	.0647	212.743	15.456	13.141	203.109	**45**
50	18.420	.0543	.00344	.0634	290.335	15.762	13.796	217.457	**50**
55	24.650	.0406	.00254	.0625	394.171	15.991	14.341	229.322	**55**
60	32.988	.0303	.00188	.0619	533.126	16.161	14.791	239.043	**60**
65	44.145	.0227	.00139	.0614	719.080	16.289	15.160	246.945	**65**
70	59.076	.0169	.00103	.0610	967.928	16.385	15.461	253.327	**70**
75	79.057	.0126	.00077	.0608	1 300.9	16.456	15.706	258.453	**75**
80	105.796	.00945	.00057	.0606	1 746.6	16.509	15.903	262.549	**80**
85	141.578	.00706	.00043	.0604	2 343.0	16.549	16.062	265.810	**85**
90	189.464	.00528	.00032	.0603	3 141.1	16.579	16.189	268.395	**90**
95	253.545	.00394	.00024	.0602	4 209.1	16.601	16.290	270.437	**95**
100	339.300	.00295	.00018	.0602	5 638.3	16.618	16.371	272.047	**100**

7% **Compound Interest Factors** 7%

	Single Payment		Uniform Payment Series				Arithmetic Gradient		
n	Compound Amount Factor Find F Given P F/P	Present Worth Factor Find P Given F P/F	Sinking Fund Factor Find A Given F A/F	Capital Recovery Factor Find A Given P A/P	Compound Amount Factor Find F Given A F/A	Present Worth Factor Find P Given A P/A	Gradient Uniform Series Find A Given G A/G	Gradient Present Worth Find P Given G P/G	n
1	1.070	.9346	1.0000	1.0700	1.000	0.935	0	0	1
2	1.145	.8734	.4831	.5531	2.070	1.808	0.483	0.873	2
3	1.225	.8163	.3111	.3811	3.215	2.624	0.955	2.506	3
4	1.311	.7629	.2252	.2952	4.440	3.387	1.416	4.795	4
5	1.403	.7130	.1739	.2439	5.751	4.100	1.865	7.647	5
6	1.501	.6663	.1398	.2098	7.153	4.767	2.303	10.978	6
7	1.606	.6227	.1156	.1856	8.654	5.389	2.730	14.715	7
8	1.718	.5820	.0975	.1675	10.260	5.971	3.147	18.789	8
9	1.838	.5439	.0835	.1535	11.978	6.515	3.552	23.140	9
10	1.967	.5083	.0724	.1424	13.816	7.024	3.946	27.716	10
11	2.105	.4751	.0634	.1334	15.784	7.499	4.330	32.467	11
12	2.252	.4440	.0559	.1259	17.888	7.943	4.703	37.351	12
13	2.410	.4150	.0497	.1197	20.141	8.358	5.065	42.330	13
14	2.579	.3878	.0443	.1143	22.551	8.745	5.417	47.372	14
15	2.759	.3624	.0398	.1098	25.129	9.108	5.758	52.446	15
16	2.952	.3387	.0359	.1059	27.888	9.447	6.090	57.527	16
17	3.159	.3166	.0324	.1024	30.840	9.763	6.411	62.592	17
18	3.380	.2959	.0294	.0994	33.999	10.059	6.722	67.622	18
19	3.617	.2765	.0268	.0968	37.379	10.336	7.024	72.599	19
20	3.870	.2584	.0244	.0944	40.996	10.594	7.316	77.509	20
21	4.141	.2415	.0223	.0923	44.865	10.836	7.599	82.339	21
22	4.430	.2257	.0204	.0904	49.006	11.061	7.872	87.079	22
23	4.741	.2109	.0187	.0887	53.436	11.272	8.137	91.720	23
24	5.072	.1971	.0172	.0872	58.177	11.469	8.392	96.255	24
25	5.427	.1842	.0158	.0858	63.249	11.654	8.639	100.677	25
26	5.807	.1722	.0146	.0846	68.677	11.826	8.877	104.981	26
27	6.214	.1609	.0134	.0834	74.484	11.987	9.107	109.166	27
28	6.649	.1504	.0124	.0824	80.698	12.137	9.329	113.227	28
29	7.114	.1406	.0114	.0814	87.347	12.278	9.543	117.162	29
30	7.612	.1314	.0106	.0806	94.461	12.409	9.749	120.972	30
31	8.145	.1228	.00980	.0798	102.073	12.532	9.947	124.655	31
32	8.715	.1147	.00907	.0791	110.218	12.647	10.138	128.212	32
33	9.325	.1072	.00841	.0784	118.934	12.754	10.322	131.644	33
34	9.978	.1002	.00780	.0778	128.259	12.854	10.499	134.951	34
35	10.677	.0937	.00723	.0772	138.237	12.948	10.669	138.135	35
40	14.974	.0668	.00501	.0750	199.636	13.332	11.423	152.293	40
45	21.002	.0476	.00350	.0735	285.750	13.606	12.036	163.756	45
50	29.457	.0339	.00246	.0725	406.530	13.801	12.529	172.905	50
55	41.315	.0242	.00174	.0717	575.930	13.940	12.921	180.124	55
60	57.947	.0173	.00123	.0712	813.523	14.039	13.232	185.768	60
65	81.273	.0123	.00087	.0709	1 146.8	14.110	13.476	190.145	65
70	113.990	.00877	.00062	.0706	1 614.1	14.160	13.666	193.519	70
75	159.877	.00625	.00044	.0704	2 269.7	14.196	13.814	196.104	75
80	224.235	.00446	.00031	.0703	3 189.1	14.222	13.927	198.075	80
85	314.502	.00318	.00022	.0702	4 478.6	14.240	14.015	199.572	85
90	441.105	.00227	.00016	.0702	6 287.2	14.253	14.081	200.704	90
95	618.673	.00162	.00011	.0701	8 823.9	14.263	14.132	201.558	95
100	867.720	.00115	.00008	.0701	12 381.7	14.269	14.170	202.200	100

8% **Compound Interest Factors** 8%

	Single Payment		Uniform Payment Series				Arithmetic Gradient		
	Compound Amount Factor	Present Worth Factor	Sinking Fund Factor	Capital Recovery Factor	Compound Amount Factor	Present Worth Factor	Gradient Uniform Series	Gradient Present Worth	
	Find *F* Given *P*	Find *P* Given *F*	Find *A* Given *F*	Find *A* Given *P*	Find *F* Given *A*	Find *P* Given *A*	Find *A* Given *G*	Find *P* Given *G*	
n	*F/P*	*P/F*	*A/F*	*A/P*	*F/A*	*P/A*	*A/G*	*P/G*	*n*
1	1.080	.9259	1.0000	1.0800	1.000	0.926	0	0	**1**
2	1.166	.8573	.4808	.5608	2.080	1.783	0.481	0.857	**2**
3	1.260	.7938	.3080	.3880	3.246	2.577	0.949	2.445	**3**
4	1.360	.7350	.2219	.3019	4.506	3.312	1.404	4.650	**4**
5	1.469	.6806	.1705	.2505	5.867	3.993	1.846	7.372	**5**
6	1.587	.6302	.1363	.2163	7.336	4.623	2.276	10.523	**6**
7	1.714	.5835	.1121	.1921	8.923	5.206	2.694	14.024	**7**
8	1.851	.5403	.0940	.1740	10.637	5.747	3.099	17.806	**8**
9	1.999	.5002	.0801	.1601	12.488	6.247	3.491	21.808	**9**
10	2.159	.4632	.0690	.1490	14.487	6.710	3.871	25.977	**10**
11	2.332	.4289	.0601	.1401	16.645	7.139	4.240	30.266	**11**
12	2.518	.3971	.0527	.1327	18.977	7.536	4.596	34.634	**12**
13	2.720	.3677	.0465	.1265	21.495	7.904	4.940	39.046	**13**
14	2.937	.3405	.0413	.1213	24.215	8.244	5.273	43.472	**14**
15	3.172	.3152	.0368	.1168	27.152	8.559	5.594	47.886	**15**
16	3.426	.2919	.0330	.1130	30.324	8.851	5.905	52.264	**16**
17	3.700	.2703	.0296	.1096	33.750	9.122	6.204	56.588	**17**
18	3.996	.2502	.0267	.1067	37.450	9.372	6.492	60.843	**18**
19	4.316	.2317	.0241	.1041	41.446	9.604	6.770	65.013	**19**
20	4.661	.2145	.0219	.1019	45.762	9.818	7.037	69.090	**20**
21	5.034	.1987	.0198	.0998	50.423	10.017	7.294	73.063	**21**
22	5.437	.1839	.0180	.0980	55.457	10.201	7.541	76.926	**22**
23	5.871	.1703	.0164	.0964	60.893	10.371	7.779	80.673	**23**
24	6.341	.1577	.0150	.0950	66.765	10.529	8.007	84.300	**24**
25	6.848	.1460	.0137	.0937	73.106	10.675	8.225	87.804	**25**
26	7.396	.1352	.0125	.0925	79.954	10.810	8.435	91.184	**26**
27	7.988	.1252	.0114	.0914	87.351	10.935	8.636	94.439	**27**
28	8.627	.1159	.0105	.0905	95.339	11.051	8.829	97.569	**28**
29	9.317	.1073	.00962	.0896	103.966	11.158	9.013	100.574	**29**
30	10.063	.0994	.00883	.0888	113.283	11.258	9.190	103.456	**30**
31	10.868	.0920	.00811	.0881	123.346	11.350	9.358	106.216	**31**
32	11.737	.0852	.00745	.0875	134.214	11.435	9.520	108.858	**32**
33	12.676	.0789	.00685	.0869	145.951	11.514	9.674	111.382	**33**
34	13.690	.0730	.00630	.0863	158.627	11.587	9.821	113.792	**34**
35	14.785	.0676	.00580	.0858	172.317	11.655	9.961	116.092	**35**
40	21.725	.0460	.00386	.0839	259.057	11.925	10.570	126.042	**40**
45	31.920	.0313	.00259	.0826	386.506	12.108	11.045	133.733	**45**
50	46.902	.0213	.00174	.0817	573.771	12.233	11.411	139.593	**50**
55	68.914	.0145	.00118	.0812	848.925	12.319	11.690	144.006	**55**
60	101.257	.00988	.00080	.0808	1 253.2	12.377	11.902	147.300	**60**
65	148.780	.00672	.00054	.0805	1 847.3	12.416	12.060	149.739	**65**
70	218.607	.00457	.00037	.0804	2 720.1	12.443	12.178	151.533	**70**
75	321.205	.00311	.00025	.0802	4 002.6	12.461	12.266	152.845	**75**
80	471.956	.00212	.00017	.0802	5 887.0	12.474	12.330	153.800	**80**
85	693.458	.00144	.00012	.0801	8 655.7	12.482	12.377	154.492	**85**
90	1 018.9	.00098	.00008	.0801	12 724.0	12.488	12.412	154.993	**90**
95	1 497.1	.00067	.00005	.0801	18 701.6	12.492	12.437	155.352	**95**
100	2 199.8	.00045	.00004	.0800	27 484.6	12.494	12.455	155.611	**100**

9% **Compound Interest Factors** **9%**

	Single Payment		Uniform Payment Series				Arithmetic Gradient		
n	Compound Amount Factor Find F Given P F/P	Present Worth Factor Find P Given F P/F	Sinking Fund Factor Find A Given F A/F	Capital Recovery Factor Find A Given P A/P	Compound Amount Factor Find F Given A F/A	Present Worth Factor Find P Given A P/A	Gradient Uniform Series Find A Given G A/G	Gradient Present Worth Find P Given G P/G	n
1	1.090	.9174	1.0000	1.0900	1.000	0.917	0	0	**1**
2	1.188	.8417	.4785	.5685	2.090	1.759	0.478	0.842	**2**
3	1.295	.7722	.3051	.3951	3.278	2.531	0.943	2.386	**3**
4	1.412	.7084	.2187	.3087	4.573	3.240	1.393	4.511	**4**
5	1.539	.6499	.1671	.2571	5.985	3.890	1.828	7.111	**5**
6	1.677	.5963	.1329	.2229	7.523	4.486	2.250	10.092	**6**
7	1.828	.5470	.1087	.1987	9.200	5.033	2.657	13.375	**7**
8	1.993	.5019	.0907	.1807	11.028	5.535	3.051	16.888	**8**
9	2.172	.4604	.0768	.1668	13.021	5.995	3.431	20.571	**9**
10	2.367	.4224	.0658	.1558	15.193	6.418	3.798	24.373	**10**
11	2.580	.3875	.0569	.1469	17.560	6.805	4.151	28.248	**11**
12	2.813	.3555	.0497	.1397	20.141	7.161	4.491	32.159	**12**
13	3.066	.3262	.0436	.1336	22.953	7.487	4.818	36.073	**13**
14	3.342	.2992	.0384	.1284	26.019	7.786	5.133	39.963	**14**
15	3.642	.2745	.0341	.1241	29.361	8.061	5.435	43.807	**15**
16	3.970	.2519	.0303	.1203	33.003	8.313	5.724	47.585	**16**
17	4.328	.2311	.0270	.1170	36.974	8.544	6.002	51.282	**17**
18	4.717	.2120	.0242	.1142	41.301	8.756	6.269	54.886	**18**
19	5.142	.1945	.0217	.1117	46.019	8.950	6.524	58.387	**19**
20	5.604	.1784	.0195	.1095	51.160	9.129	6.767	61.777	**20**
21	6.109	.1637	.0176	.1076	56.765	9.292	7.001	65.051	**21**
22	6.659	.1502	.0159	.1059	62.873	9.442	7.223	68.205	**22**
23	7.258	.1378	.0144	.1044	69.532	9.580	7.436	71.236	**23**
24	7.911	.1264	.0130	.1030	76.790	9.707	7.638	74.143	**24**
25	8.623	.1160	.0118	.1018	84.701	9.823	7.832	76.927	**25**
26	9.399	.1064	.0107	.1007	93.324	9.929	8.016	79.586	**26**
27	10.245	.0976	.00973	.0997	102.723	10.027	8.191	82.124	**27**
28	11.167	.0895	.00885	.0989	112.968	10.116	8.357	84.542	**28**
29	12.172	.0822	.00806	.0981	124.136	10.198	8.515	86.842	**29**
30	13.268	.0754	.00734	.0973	136.308	10.274	8.666	89.028	**30**
31	14.462	.0691	.00669	.0967	149.575	10.343	8.808	91.102	**31**
32	15.763	.0634	.00610	.0961	164.037	10.406	8.944	93.069	**32**
33	17.182	.0582	.00556	.0956	179.801	10.464	9.072	94.931	**33**
34	18.728	.0534	.00508	.0951	196.983	10.518	9.193	96.693	**34**
35	20.414	.0490	.00464	.0946	215.711	10.567	9.308	98.359	**35**
40	31.409	.0318	.00296	.0930	337.883	10.757	9.796	105.376	**40**
45	48.327	.0207	.00190	.0919	525.860	10.881	10.160	110.556	**45**
50	74.358	.0134	.00123	.0912	815.085	10.962	10.430	114.325	**50**
55	114.409	.00874	.00079	.0908	1 260.1	11.014	10.626	117.036	**55**
60	176.032	.00568	.00051	.0905	1 944.8	11.048	10.768	118.968	**60**
65	270.847	.00369	.00033	.0903	2 998.3	11.070	10.870	120.334	**65**
70	416.731	.00240	.00022	.0902	4 619.2	11.084	10.943	121.294	**70**
75	641.193	.00156	.00014	.0901	7 113.3	11.094	10.994	121.965	**75**
80	986.555	.00101	.00009	.0901	10 950.6	11.100	11.030	122.431	**80**
85	1 517.9	.00066	.00006	.0901	16 854.9	11.104	11.055	122.753	**85**
90	2 335.5	.00043	.00004	.0900	25 939.3	11.106	11.073	122.976	**90**
95	3 593.5	.00028	.00003	.0900	39 916.8	11.108	11.085	123.129	**95**
100	5 529.1	.00018	.00002	.0900	61 422.9	11.109	11.093	123.233	**100**

10% **Compound Interest Factors** **10%**

	Single Payment		Uniform Payment Series				Arithmetic Gradient		
n	Compound Amount Factor Find *F* Given *P* *F/P*	Present Worth Factor Find *P* Given *F* *P/F*	Sinking Fund Factor Find *A* Given *F* *A/F*	Capital Recovery Factor Find *A* Given *P* *A/P*	Compound Amount Factor Find *F* Given *A* *F/A*	Present Worth Factor Find *P* Given *A* *P/A*	Gradient Uniform Series Find *A* Given *G* *A/G*	Gradient Present Worth Find *P* Given *G* *P/G*	n
1	1.100	.9091	1.0000	1.1000	1.000	0.909	0	0	**1**
2	1.210	.8264	.4762	.5762	2.100	1.736	0.476	0.826	**2**
3	1.331	.7513	.3021	.4021	3.310	2.487	0.937	2.329	**3**
4	1.464	.6830	.2155	.3155	4.641	3.170	1.381	4.378	**4**
5	1.611	.6209	.1638	.2638	6.105	3.791	1.810	6.862	**5**
6	1.772	.5645	.1296	.2296	7.716	4.355	2.224	9.684	**6**
7	1.949	.5132	.1054	.2054	9.487	4.868	2.622	12.763	**7**
8	2.144	.4665	.0874	.1874	11.436	5.335	3.004	16.029	**8**
9	2.358	.4241	.0736	.1736	13.579	5.759	3.372	19.421	**9**
10	2.594	.3855	.0627	.1627	15.937	6.145	3.725	22.891	**10**
11	2.853	.3505	.0540	.1540	18.531	6.495	4.064	26.396	**11**
12	3.138	.3186	.0468	.1468	21.384	6.814	4.388	29.901	**12**
13	3.452	.2897	.0408	.1408	24.523	7.103	4.699	33.377	**13**
14	3.797	.2633	.0357	.1357	27.975	7.367	4.996	36.801	**14**
15	4.177	.2394	.0315	.1315	31.772	7.606	5.279	40.152	**15**
16	4.595	.2176	.0278	.1278	35.950	7.824	5.549	43.416	**16**
17	5.054	.1978	.0247	.1247	40.545	8.022	5.807	46.582	**17**
18	5.560	.1799	.0219	.1219	45.599	8.201	6.053	49.640	**18**
19	6.116	.1635	.0195	.1195	51.159	8.365	6.286	52.583	**19**
20	6.728	.1486	.0175	.1175	57.275	8.514	6.508	55.407	**20**
21	7.400	.1351	.0156	.1156	64.003	8.649	6.719	58.110	**21**
22	8.140	.1228	.0140	.1140	71.403	8.772	6.919	60.689	**22**
23	8.954	.1117	.0126	.1126	79.543	8.883	7.108	63.146	**23**
24	9.850	.1015	.0113	.1113	88.497	8.985	7.288	65.481	**24**
25	10.835	.0923	.0102	.1102	98.347	9.077	7.458	67.696	**25**
26	11.918	.0839	.00916	.1092	109.182	9.161	7.619	69.794	**26**
27	13.110	.0763	.00826	.1083	121.100	9.237	7.770	71.777	**27**
28	14.421	.0693	.00745	.1075	134.210	9.307	7.914	73.650	**28**
29	15.863	.0630	.00673	.1067	148.631	9.370	8.049	75.415	**29**
30	17.449	.0573	.00608	.1061	164.494	9.427	8.176	77.077	**30**
31	19.194	.0521	.00550	.1055	181.944	9.479	8.296	78.640	**31**
32	21.114	.0474	.00497	.1050	201.138	9.526	8.409	80.108	**32**
33	23.225	.0431	.00450	.1045	222.252	9.569	8.515	81.486	**33**
34	25.548	.0391	.00407	.1041	245.477	9.609	8.615	82.777	**34**
35	28.102	.0356	.00369	.1037	271.025	9.644	8.709	83.987	**35**
40	45.259	.0221	.00226	.1023	442.593	9.779	9.096	88.953	**40**
45	72.891	.0137	.00139	.1014	718.905	9.863	9.374	92.454	**45**
50	117.391	.00852	.00086	.1009	1 163.9	9.915	9.570	94.889	**50**
55	189.059	.00529	.00053	.1005	1 880.6	9.947	9.708	96.562	**55**
60	304.482	.00328	.00033	.1003	3 034.8	9.967	9.802	97.701	**60**
65	490.371	.00204	.00020	.1002	4 893.7	9.980	9.867	98.471	**65**
70	789.748	.00127	.00013	.1001	7 887.5	9.987	9.911	98.987	**70**
75	1 271.9	.00079	.00008	.1001	12 709.0	9.992	9.941	99.332	**75**
80	2 048.4	.00049	.00005	.1000	20 474.0	9.995	9.961	99.561	**80**
85	3 299.0	.00030	.00003	.1000	32 979.7	9.997	9.974	99.712	**85**
90	5 313.0	.00019	.00002	.1000	53 120.3	9.998	9.983	99.812	**90**
95	8 556.7	.00012	.00001	.1000	85 556.9	9.999	9.989	99.877	**95**
100	13 780.6	.00007	.00001	.1000	137 796.3	9.999	9.993	99.920	**100**

12% **Compound Interest Factors** 12%

	Single Payment		Uniform Payment Series				Arithmetic Gradient		
	Compound Amount Factor	Present Worth Factor	Sinking Fund Factor	Capital Recovery Factor	Compound Amount Factor	Present Worth Factor	Gradient Uniform Series	Gradient Present Worth	
n	Find *F* Given *P* *F*/*P*	Find *P* Given *F* *P*/*F*	Find *A* Given *F* *A*/*F*	Find *A* Given *P* *A*/*P*	Find *F* Given *A* *F*/*A*	Find *P* Given *A* *P*/*A*	Find *A* Given *G* *A*/*G*	Find *P* Given *G* *P*/*G*	*n*
1	1.120	.8929	1.0000	1.1200	1.000	0.893	0	0	**1**
2	1.254	.7972	.4717	.5917	2.120	1.690	0.472	0.797	**2**
3	1.405	.7118	.2963	.4163	3.374	2.402	0.925	2.221	**3**
4	1.574	.6355	.2092	.3292	4.779	3.037	1.359	4.127	**4**
5	1.762	.5674	.1574	.2774	6.353	3.605	1.775	6.397	**5**
6	1.974	.5066	.1232	.2432	8.115	4.111	2.172	8.930	**6**
7	2.211	.4523	.0991	.2191	10.089	4.564	2.551	11.644	**7**
8	2.476	.4039	.0813	.2013	12.300	4.968	2.913	14.471	**8**
9	2.773	.3606	.0677	.1877	14.776	5.328	3.257	17.356	**9**
10	3.106	.3220	.0570	.1770	17.549	5.650	3.585	20.254	**10**
11	3.479	.2875	.0484	.1684	20.655	5.938	3.895	23.129	**11**
12	3.896	.2567	.0414	.1614	24.133	6.194	4.190	25.952	**12**
13	4.363	.2292	.0357	.1557	28.029	6.424	4.468	28.702	**13**
14	4.887	.2046	.0309	.1509	32.393	6.628	4.732	31.362	**14**
15	5.474	.1827	.0268	.1468	37.280	6.811	4.980	33.920	**15**
16	6.130	.1631	.0234	.1434	42.753	6.974	5.215	36.367	**16**
17	6.866	.1456	.0205	.1405	48.884	7.120	5.435	38.697	**17**
18	7.690	.1300	.0179	.1379	55.750	7.250	5.643	40.908	**18**
19	8.613	.1161	.0158	.1358	63.440	7.366	5.838	42.998	**19**
20	9.646	.1037	.0139	.1339	72.052	7.469	6.020	44.968	**20**
21	10.804	.0926	.0122	.1322	81.699	7.562	6.191	46.819	**21**
22	12.100	.0826	.0108	.1308	92.503	7.645	6.351	48.554	**22**
23	13.552	.0738	.00956	.1296	104.603	7.718	6.501	50.178	**23**
24	15.179	.0659	.00846	.1285	118.155	7.784	6.641	51.693	**24**
25	17.000	.0588	.00750	.1275	133.334	7.843	6.771	53.105	**25**
26	19.040	.0525	.00665	.1267	150.334	7.896	6.892	54.418	**26**
27	21.325	.0469	.00590	.1259	169.374	7.943	7.005	55.637	**27**
28	23.884	.0419	.00524	.1252	190.699	7.984	7.110	56.767	**28**
29	26.750	.0374	.00466	.1247	214.583	8.022	7.207	57.814	**29**
30	29.960	.0334	.00414	.1241	241.333	8.055	7.297	58.782	**30**
31	33.555	.0298	.00369	.1237	271.293	8.085	7.381	59.676	**31**
32	37.582	.0266	.00328	.1233	304.848	8.112	7.459	60.501	**32**
33	42.092	.0238	.00292	.1229	342.429	8.135	7.530	61.261	**33**
34	47.143	.0212	.00260	.1226	384.521	8.157	7.596	61.961	**34**
35	52.800	.0189	.00232	.1223	431.663	8.176	7.658	62.605	**35**
40	93.051	.0107	.00130	.1213	767.091	8.244	7.899	65.116	**40**
45	163.988	.00610	.00074	.1207	1 358.2	8.283	8.057	66.734	**45**
50	289.002	.00346	.00042	.1204	2 400.0	8.304	8.160	67.762	**50**
55	509.321	.00196	.00024	.1202	4 236.0	8.317	8.225	68.408	**55**
60	897.597	.00111	.00013	.1201	7 471.6	8.324	8.266	68.810	**60**
65	1 581.9	.00063	.00008	.1201	13 173.9	8.328	8.292	69.058	**65**
70	2 787.8	.00036	.00004	.1200	23 223.3	8.330	8.308	69.210	**70**
75	4 913.1	.00020	.00002	.1200	40 933.8	8.332	8.318	69.303	**75**
80	8 658.5	.00012	.00001	.1200	72 145.7	8.332	8.324	69.359	**80**
85	15 259.2	.00007	.00001	.1200	127 151.7	8.333	8.328	69.393	**85**
90	26 891.9	.00004		.1200	224 091.1	8.333	8.330	69.414	**90**
95	47 392.8	.00002		.1200	394 931.4	8.333	8.331	69.426	**95**
100	83 522.3	.00001		.1200	696 010.5	8.333	8.332	69.434	**100**

15% **Compound Interest Factors** 15%

	Single Payment		Uniform Payment Series				Arithmetic Gradient		
	Compound Amount Factor	Present Worth Factor	Sinking Fund Factor	Capital Recovery Factor	Compound Amount Factor	Present Worth Factor	Gradient Uniform Series	Gradient Present Worth	
	Find *F* Given *P*	Find *P* Given *F*	Find *A* Given *F*	Find *A* Given *P*	Find *F* Given *A*	Find *P* Given *A*	Find *A* Given *G*	Find *P* Given *G*	
n	*F/P*	*P/F*	*A/F*	*A/P*	*F/A*	*P/A*	*A/G*	*P/G*	*n*
1	1.150	.8696	1.0000	1.1500	1.000	0.870	0	0	**1**
2	1.322	.7561	.4651	.6151	2.150	1.626	0.465	0.756	**2**
3	1.521	.6575	.2880	.4380	3.472	2.283	0.907	2.071	**3**
4	1.749	.5718	.2003	.3503	4.993	2.855	1.326	3.786	**4**
5	2.011	.4972	.1483	.2983	6.742	3.352	1.723	5.775	**5**
6	2.313	.4323	.1142	.2642	8.754	3.784	2.097	7.937	**6**
7	2.660	.3759	.0904	.2404	11.067	4.160	2.450	10.192	**7**
8	3.059	.3269	.0729	.2229	13.727	4.487	2.781	12.481	**8**
9	3.518	.2843	.0596	.2096	16.786	4.772	3.092	14.755	**9**
10	4.046	.2472	.0493	.1993	20.304	5.019	3.383	16.979	**10**
11	4.652	.2149	.0411	.1911	24.349	5.234	3.655	19.129	**11**
12	5.350	.1869	.0345	.1845	29.002	5.421	3.908	21.185	**12**
13	6.153	.1625	.0291	.1791	34.352	5.583	4.144	23.135	**13**
14	7.076	.1413	.0247	.1747	40.505	5.724	4.362	24.972	**14**
15	8.137	.1229	.0210	.1710	47.580	5.847	4.565	26.693	**15**
16	9.358	.1069	.0179	.1679	55.717	5.954	4.752	28.296	**16**
17	10.761	.0929	.0154	.1654	65.075	6.047	4.925	29.783	**17**
18	12.375	.0808	.0132	.1632	75.836	6.128	5.084	31.156	**18**
19	14.232	.0703	.0113	.1613	88.212	6.198	5.231	32.421	**19**
20	16.367	.0611	.00976	.1598	102.444	6.259	5.365	33.582	**20**
21	18.822	.0531	.00842	.1584	118.810	6.312	5.488	34.645	**21**
22	21.645	.0462	.00727	.1573	137.632	6.359	5.601	35.615	**22**
23	24.891	.0402	.00628	.1563	159.276	6.399	5.704	36.499	**23**
24	28.625	.0349	.00543	.1554	184.168	6.434	5.798	37.302	**24**
25	32.919	.0304	.00470	.1547	212.793	6.464	5.883	38.031	**25**
26	37.857	.0264	.00407	.1541	245.712	6.491	5.961	38.692	**26**
27	43.535	.0230	.00353	.1535	283.569	6.514	6.032	39.289	**27**
28	50.066	.0200	.00306	.1531	327.104	6.534	6.096	39.828	**28**
29	57.575	.0174	.00265	.1527	377.170	6.551	6.154	40.315	**29**
30	66.212	.0151	.00230	.1523	434.745	6.566	6.207	40.753	**30**
31	76.144	.0131	.00200	.1520	500.957	6.579	6.254	41.147	**31**
32	87.565	.0114	.00173	.1517	577.100	6.591	6.297	41.501	**32**
33	100.700	.00993	.00150	.1515	664.666	6.600	6.336	41.818	**33**
34	115.805	.00864	.00131	.1513	765.365	6.609	6.371	42.103	**34**
35	133.176	.00751	.00113	.1511	881.170	6.617	6.402	42.359	**35**
40	267.864	.00373	.00056	.1506	1 779.1	6.642	6.517	43.283	**40**
45	538.769	.00186	.00028	.1503	3 585.1	6.654	6.583	43.805	**45**
50	1 083.7	.00092	.00014	.1501	7 217.7	6.661	6.620	44.096	**50**
55	2 179.6	.00046	.00007	.1501	14 524.1	6.664	6.641	44.256	**55**
60	4 384.0	.00023	.00003	.1500	29 220.0	6.665	6.653	44.343	**60**
65	8 817.8	.00011	.00002	.1500	58 778.6	6.666	6.659	44.390	**65**
70	17 735.7	.00006	.00001	.1500	118 231.5	6.666	6.663	44.416	**70**
75	35 672.9	.00003		.1500	237 812.5	6.666	6.665	44.429	**75**
80	71 750.9	.00001		.1500	478 332.6	6.667	6.666	44.436	**80**
85	144 316.7	.00001		.1500	962 104.4	6.667	6.666	44.440	**85**

18% **Compound Interest Factors** 18%

	Single Payment		Uniform Payment Series				Arithmetic Gradient		
	Compound Amount Factor	Present Worth Factor	Sinking Fund Factor	Capital Recovery Factor	Compound Amount Factor	Present Worth Factor	Gradient Uniform Series	Gradient Present Worth	
	Find *F* Given *P*	Find *P* Given *F*	Find *A* Given *F*	Find *A* Given *P*	Find *F* Given *A*	Find *P* Given *A*	Find *A* Given *G*	Find *P* Given *G*	
n	*F/P*	*P/F*	*A/F*	*A/P*	*F/A*	*P/A*	*A/G*	*P/G*	*n*
1	1.180	.8475	1.0000	1.1800	1.000	0.847	0	0	**1**
2	1.392	.7182	.4587	.6387	2.180	1.566	0.459	0.718	**2**
3	1.643	.6086	.2799	.4599	3.572	2.174	0.890	1.935	**3**
4	1.939	.5158	.1917	.3717	5.215	2.690	1.295	3.483	**4**
5	2.288	.4371	.1398	.3198	7.154	3.127	1.673	5.231	**5**
6	2.700	.3704	.1059	.2859	9.442	3.498	2.025	7.083	**6**
7	3.185	.3139	.0824	.2624	12.142	3.812	2.353	8.967	**7**
8	3.759	.2660	.0652	.2452	15.327	4.078	2.656	10.829	**8**
9	4.435	.2255	.0524	.2324	19.086	4.303	2.936	12.633	**9**
10	5.234	.1911	.0425	.2225	23.521	4.494	3.194	14.352	**10**
11	6.176	.1619	.0348	.2148	28.755	4.656	3.430	15.972	**11**
12	7.288	.1372	.0286	.2086	34.931	4.793	3.647	17.481	**12**
13	8.599	.1163	.0237	.2037	42.219	4.910	3.845	18.877	**13**
14	10.147	.0985	.0197	.1997	50.818	5.008	4.025	20.158	**14**
15	11.974	.0835	.0164	.1964	60.965	5.092	4.189	21.327	**15**
16	14.129	.0708	.0137	.1937	72.939	5.162	4.337	22.389	**16**
17	16.672	.0600	.0115	.1915	87.068	5.222	4.471	23.348	**17**
18	19.673	.0508	.00964	.1896	103.740	5.273	4.592	24.212	**18**
19	23.214	.0431	.00810	.1881	123.413	5.316	4.700	24.988	**19**
20	27.393	.0365	.00682	.1868	146.628	5.353	4.798	25.681	**20**
21	32.324	.0309	.00575	.1857	174.021	5.384	4.885	26.300	**21**
22	38.142	.0262	.00485	.1848	206.345	5.410	4.963	26.851	**22**
23	45.008	.0222	.00409	.1841	244.487	5.432	5.033	27.339	**23**
24	53.109	.0188	.00345	.1835	289.494	5.451	5.095	27.772	**24**
25	62.669	.0160	.00292	.1829	342.603	5.467	5.150	28.155	**25**
26	73.949	.0135	.00247	.1825	405.272	5.480	5.199	28.494	**26**
27	87.260	.0115	.00209	.1821	479.221	5.492	5.243	28.791	**27**
28	102.966	.00971	.00177	.1818	566.480	5.502	5.281	29.054	**28**
29	121.500	.00823	.00149	.1815	669.447	5.510	5.315	29.284	**29**
30	143.370	.00697	.00126	.1813	790.947	5.517	5.345	29.486	**30**
31	169.177	.00591	.00107	.1811	934.317	5.523	5.371	29.664	**31**
32	199.629	.00501	.00091	.1809	1 103.5	5.528	5.394	29.819	**32**
33	235.562	.00425	.00077	.1808	1 303.1	5.532	5.415	29.955	**33**
34	277.963	.00360	.00065	.1806	1 538.7	5.536	5.433	30.074	**34**
35	327.997	.00305	.00055	.1806	1 816.6	5.539	5.449	30.177	**35**
40	750.377	.00133	.00024	.1802	4 163.2	5.548	5.502	30.527	**40**
45	1 716.7	.00058	.00010	.1801	9 531.6	5.552	5.529	30.701	**45**
50	3 927.3	.00025	.00005	.1800	21 813.0	5.554	5.543	30.786	**50**
55	8 984.8	.00011	.00002	.1800	49 910.1	5.555	5.549	30.827	**55**
60	20 555.1	.00005	.00001	.1800	114 189.4	5.555	5.553	30.846	**60**
65	47 025.1	.00002		.1800	261 244.7	5.555	5.554	30.856	**65**
70	107 581.9	.00001		.1800	597 671.7	5.556	5.555	30.860	**70**

20% **Compound Interest Factors** **20%**

	Single Payment		Uniform Payment Series				Arithmetic Gradient		
	Compound Amount Factor	Present Worth Factor	Sinking Fund Factor	Capital Recovery Factor	Compound Amount Factor	Present Worth Factor	Gradient Uniform Series	Gradient Present Worth	
n	Find *F* Given *P* *F/P*	Find *P* Given *F* *P/F*	Find *A* Given *F* *A/F*	Find *A* Given *P* *A/P*	Find *F* Given *A* *F/A*	Find *P* Given *A* *P/A*	Find *A* Given *G* *A/G*	Find *P* Given *G* *P/G*	*n*
1	1.200	.8333	1.0000	1.2000	1.000	0.833	0	0	**1**
2	1.440	.6944	.4545	.6545	2.200	1.528	0.455	0.694	**2**
3	1.728	.5787	.2747	.4747	3.640	2.106	0.879	1.852	**3**
4	2.074	.4823	.1863	.3863	5.368	2.589	1.274	3.299	**4**
5	2.488	.4019	.1344	.3344	7.442	2.991	1.641	4.906	**5**
6	2.986	.3349	.1007	.3007	9.930	3.326	1.979	6.581	**6**
7	3.583	.2791	.0774	.2774	12.916	3.605	2.290	8.255	**7**
8	4.300	.2326	.0606	.2606	16.499	3.837	2.576	9.883	**8**
9	5.160	.1938	.0481	.2481	20.799	4.031	2.836	11.434	**9**
10	6.192	.1615	.0385	.2385	25.959	4.192	3.074	12.887	**10**
11	7.430	.1346	.0311	.2311	32.150	4.327	3.289	14.233	**11**
12	8.916	.1122	.0253	.2253	39.581	4.439	3.484	15.467	**12**
13	10.699	.0935	.0206	.2206	48.497	4.533	3.660	16.588	**13**
14	12.839	.0779	.0169	.2169	59.196	4.611	3.817	17.601	**14**
15	15.407	.0649	.0139	.2139	72.035	4.675	3.959	18.509	**15**
16	18.488	.0541	.0114	.2114	87.442	4.730	4.085	19.321	**16**
17	22.186	.0451	.00944	.2094	105.931	4.775	4.198	20.042	**17**
18	26.623	.0376	.00781	.2078	128.117	4.812	4.298	20.680	**18**
19	31.948	.0313	.00646	.2065	154.740	4.843	4.386	21.244	**19**
20	38.338	.0261	.00536	.2054	186.688	4.870	4.464	21.739	**20**
21	46.005	.0217	.00444	.2044	225.026	4.891	4.533	22.174	**21**
22	55.206	.0181	.00369	.2037	271.031	4.909	4.594	22.555	**22**
23	66.247	.0151	.00307	.2031	326.237	4.925	4.647	22.887	**23**
24	79.497	.0126	.00255	.2025	392.484	4.937	4.694	23.176	**24**
25	95.396	.0105	.00212	.2021	471.981	4.948	4.735	23.428	**25**
26	114.475	.00874	.00176	.2018	567.377	4.956	4.771	23.646	**26**
27	137.371	.00728	.00147	.2015	681.853	4.964	4.802	23.835	**27**
28	164.845	.00607	.00122	.2012	819.223	4.970	4.829	23.999	**28**
29	197.814	.00506	.00102	.2010	984.068	4.975	4.853	24.141	**29**
30	237.376	.00421	.00085	.2008	1 181.9	4.979	4.873	24.263	**30**
31	284.852	.00351	.00070	.2007	1 419.3	4.982	4.891	24.368	**31**
32	341.822	.00293	.00059	.2006	1 704.1	4.985	4.906	24.459	**32**
33	410.186	.00244	.00049	.2005	2 045.9	4.988	4.919	24.537	**33**
34	492.224	.00203	.00041	.2004	2 456.1	4.990	4.931	24.604	**34**
35	590.668	.00169	.00034	.2003	2 948.3	4.992	4.941	24.661	**35**
40	1 469.8	.00068	.00014	.2001	7 343.9	4.997	4.973	24.847	**40**
45	3 657.3	.00027	.00005	.2001	18 281.3	4.999	4.988	24.932	**45**
50	9 100.4	.00011	.00002	.2000	45 497.2	4.999	4.995	24.970	**50**
55	22 644.8	.00004	.00001	.2000	113 219.0	5.000	4.998	24.987	**55**
60	56 347.5	.00002		.2000	281 732.6	5.000	4.999	24.994	**60**

25% **Compound Interest Factors** 25%

	Single Payment		Uniform Payment Series				Arithmetic Gradient		
	Compound Amount Factor	Present Worth Factor	Sinking Fund Factor	Capital Recovery Factor	Compound Amount Factor	Present Worth Factor	Gradient Uniform Series	Gradient Present Worth	
n	Find *F* Given *P* *F/P*	Find *P* Given *F* *P/F*	Find *A* Given *F* *A/F*	Find *A* Given *P* *A/P*	Find *F* Given *A* *F/A*	Find *P* Given *A* *P/A*	Find *A* Given *G* *A/G*	Find *P* Given *G* *P/G*	*n*
1	1.250	.8000	1.0000	1.2500	1.000	0.800	0	0	**1**
2	1.563	.6400	.4444	.6944	2.250	1.440	0.444	0.640	**2**
3	1.953	.5120	.2623	.5123	3.813	1.952	0.852	1.664	**3**
4	2.441	.4096	.1734	.4234	5.766	2.362	1.225	2.893	**4**
5	3.052	.3277	.1218	.3718	8.207	2.689	1.563	4.204	**5**
6	3.815	.2621	.0888	.3388	11.259	2.951	1.868	5.514	**6**
7	4.768	.2097	.0663	.3163	15.073	3.161	2.142	6.773	**7**
8	5.960	.1678	.0504	.3004	19.842	3.329	2.387	7.947	**8**
9	7.451	.1342	.0388	.2888	25.802	3.463	2.605	9.021	**9**
10	9.313	.1074	.0301	.2801	33.253	3.571	2.797	9.987	**10**
11	11.642	.0859	.0235	.2735	42.566	3.656	2.966	10.846	**11**
12	14.552	.0687	.0184	.2684	54.208	3.725	3.115	11.602	**12**
13	18.190	.0550	.0145	.2645	68.760	3.780	3.244	12.262	**13**
14	22.737	.0440	.0115	.2615	86.949	3.824	3.356	12.833	**14**
15	28.422	.0352	.00912	.2591	109.687	3.859	3.453	13.326	**15**
16	35.527	.0281	.00724	.2572	138.109	3.887	3.537	13.748	**16**
17	44.409	.0225	.00576	.2558	173.636	3.910	3.608	14.108	**17**
18	55.511	.0180	.00459	.2546	218.045	3.928	3.670	14.415	**18**
19	69.389	.0144	.00366	.2537	273.556	3.942	3.722	14.674	**19**
20	86.736	.0115	.00292	.2529	342.945	3.954	3.767	14.893	**20**
21	108.420	.00922	.00233	.2523	429.681	3.963	3.805	15.078	**21**
22	135.525	.00738	.00186	.2519	538.101	3.970	3.836	15.233	**22**
23	169.407	.00590	.00148	.2515	673.626	3.976	3.863	15.362	**23**
24	211.758	.00472	.00119	.2512	843.033	3.981	3.886	15.471	**24**
25	264.698	.00378	.00095	.2509	1 054.8	3.985	3.905	15.562	**25**
26	330.872	.00302	.00076	.2508	1 319.5	3.988	3.921	15.637	**26**
27	413.590	.00242	.00061	.2506	1 650.4	3.990	3.935	15.700	**27**
28	516.988	.00193	.00048	.2505	2 064.0	3.992	3.946	15.752	**28**
29	646.235	.00155	.00039	.2504	2 580.9	3.994	3.955	15.796	**29**
30	807.794	.00124	.00031	.2503	3 227.2	3.995	3.963	15.832	**30**
31	1 009.7	.00099	.00025	.2502	4 035.0	3.996	3.969	15.861	**31**
32	1 262.2	.00079	.00020	.2502	5 044.7	3.997	3.975	15.886	**32**
33	1 577.7	.00063	.00016	.2502	6 306.9	3.997	3.979	15.906	**33**
34	1 972.2	.00051	.00013	.2501	7 884.6	3.998	3.983	15.923	**34**
35	2 465.2	.00041	.00010	.2501	9 856.8	3.998	3.986	15.937	**35**
40	7 523.2	.00013	.00003	.2500	30 088.7	3.999	3.995	15.977	**40**
45	22 958.9	.00004	.00001	.2500	91 831.5	4.000	3.998	15.991	**45**
50	70 064.9	.00001		.2500	280 255.7	4.000	3.999	15.997	**50**
55	213 821.2			.2500	855 280.7	4.000	4.000	15.999	**55**

30%				Compound Interest Factors					30%
	Single Payment		Uniform Payment Series				Arithmetic Gradient		
	Compound Amount Factor	Present Worth Factor	Sinking Fund Factor	Capital Recovery Factor	Compound Amount Factor	Present Worth Factor	Gradient Uniform Series	Gradient Present Worth	
n	Find *F* Given *P* *F/P*	Find *P* Given *F* *P/F*	Find *A* Given *F* *A/F*	Find *A* Given *P* *A/P*	Find *F* Given *A* *F/A*	Find *P* Given *A* *P/A*	Find *A* Given *G* *A/G*	Find *P* Given *G* *P/G*	*n*
1	1.300	.7692	1.0000	1.3000	1.000	0.769	0	0	**1**
2	1.690	.5917	.4348	.7348	2.300	1.361	0.435	0.592	**2**
3	2.197	.4552	.2506	.5506	3.990	1.816	0.827	1.502	**3**
4	2.856	.3501	.1616	.4616	6.187	2.166	1.178	2.552	**4**
5	3.713	.2693	.1106	.4106	9.043	2.436	1.490	3.630	**5**
6	4.827	.2072	.0784	.3784	12.756	2.643	1.765	4.666	**6**
7	6.275	.1594	.0569	.3569	17.583	2.802	2.006	5.622	**7**
8	8.157	.1226	.0419	.3419	23.858	2.925	2.216	6.480	**8**
9	10.604	.0943	.0312	.3312	32.015	3.019	2.396	7.234	**9**
10	13.786	.0725	.0235	.3235	42.619	3.092	2.551	7.887	**10**
11	17.922	.0558	.0177	.3177	56.405	3.147	2.683	8.445	**11**
12	23.298	.0429	.0135	.3135	74.327	3.190	2.795	8.917	**12**
13	30.287	.0330	.0102	.3102	97.625	3.223	2.889	9.314	**13**
14	39.374	.0254	.00782	.3078	127.912	3.249	2.969	9.644	**14**
15	51.186	.0195	.00598	.3060	167.286	3.268	3.034	9.917	**15**
16	66.542	.0150	.00458	.3046	218.472	3.283	3.089	10.143	**16**
17	86.504	.0116	.00351	.3035	285.014	3.295	3.135	10.328	**17**
18	112.455	.00889	.00269	.3027	371.518	3.304	3.172	10.479	**18**
19	146.192	.00684	.00207	.3021	483.973	3.311	3.202	10.602	**19**
20	190.049	.00526	.00159	.3016	630.165	3.316	3.228	10.702	**20**
21	247.064	.00405	.00122	.3012	820.214	3.320	3.248	10.783	**21**
22	321.184	.00311	.00094	.3009	1 067.3	3.323	3.265	10.848	**22**
23	417.539	.00239	.00072	.3007	1 388.5	3.325	3.278	10.901	**23**
24	542.800	.00184	.00055	.3006	1 806.0	3.327	3.289	10.943	**24**
25	705.640	.00142	.00043	.3004	2 348.8	3.329	3.298	10.977	**25**
26	917.332	.00109	.00033	.3003	3 054.4	3.330	3.305	11.005	**26**
27	1 192.5	.00084	.00025	.3003	3 971.8	3.331	3.311	11.026	**27**
28	1 550.3	.00065	.00019	.3002	5 164.3	3.331	3.315	11.044	**28**
29	2 015.4	.00050	.00015	.3001	6 714.6	3.332	3.319	11.058	**29**
30	2 620.0	.00038	.00011	.3001	8 730.0	3.332	3.322	11.069	**30**
31	3 406.0	.00029	.00009	.3001	11 350.0	3.332	3.324	11.078	**31**
32	4 427.8	.00023	.00007	.3001	14 756.0	3.333	3.326	11.085	**32**
33	5 756.1	.00017	.00005	.3001	19 183.7	3.333	3.328	11.090	**33**
34	7 483.0	.00013	.00004	.3000	24 939.9	3.333	3.329	11.094	**34**
35	9 727.8	.00010	.00003	.3000	32 422.8	3.333	3.330	11.098	**35**
40	36 118.8	.00003	.00001	.3000	120 392.6	3.333	3.332	11.107	**40**
45	134 106.5	.00001		.3000	447 018.3	3.333	3.333	11.110	**45**

35% **Compound Interest Factors** 35%

	Single Payment		Uniform Payment Series				Arithmetic Gradient		
	Compound Amount Factor	Present Worth Factor	Sinking Fund Factor	Capital Recovery Factor	Compound Amount Factor	Present Worth Factor	Gradient Uniform Series	Gradient Present Worth	
n	Find F Given P F/P	Find P Given F P/F	Find A Given F A/F	Find A Given P A/P	Find F Given A F/A	Find P Given A P/A	Find A Given G A/G	Find P Given G P/G	n
1	1.350	.7407	1.0000	1.3500	1.000	0.741	0	0	1
2	1.822	.5487	.4255	.7755	2.350	1.289	0.426	0.549	2
3	2.460	.4064	.2397	.5897	4.173	1.696	0.803	1.362	3
4	3.322	.3011	.1508	.5008	6.633	1.997	1.134	2.265	4
5	4.484	.2230	.1005	.4505	9.954	2.220	1.422	3.157	5
6	6.053	.1652	.0693	.4193	14.438	2.385	1.670	3.983	6
7	8.172	.1224	.0488	.3988	20.492	2.508	1.881	4.717	7
8	11.032	.0906	.0349	.3849	28.664	2.598	2.060	5.352	8
9	14.894	.0671	.0252	.3752	39.696	2.665	2.209	5.889	9
10	20.107	.0497	.0183	.3683	54.590	2.715	2.334	6.336	10
11	27.144	.0368	.0134	.3634	74.697	2.752	2.436	6.705	11
12	36.644	.0273	.00982	.3598	101.841	2.779	2.520	7.005	12
13	49.470	.0202	.00722	.3572	138.485	2.799	2.589	7.247	13
14	66.784	.0150	.00532	.3553	187.954	2.814	2.644	7.442	14
15	90.158	.0111	.00393	.3539	254.739	2.825	2.689	7.597	15
16	121.714	.00822	.00290	.3529	344.897	2.834	2.725	7.721	16
17	164.314	.00609	.00214	.3521	466.611	2.840	2.753	7.818	17
18	221.824	.00451	.00158	.3516	630.925	2.844	2.776	7.895	18
19	299.462	.00334	.00117	.3512	852.748	2.848	2.793	7.955	19
20	404.274	.00247	.00087	.3509	1 152.2	2.850	2.808	8.002	20
21	545.769	.00183	.00064	.3506	1 556.5	2.852	2.819	8.038	21
22	736.789	.00136	.00048	.3505	2 102.3	2.853	2.827	8.067	22
23	994.665	.00101	.00035	.3504	2 839.0	2.854	2.834	8.089	23
24	1 342.8	.00074	.00026	.3503	3 833.7	2.855	2.839	8.106	24
25	1 812.8	.00055	.00019	.3502	5 176.5	2.856	2.843	8.119	25
26	2 447.2	.00041	.00014	.3501	6 989.3	2.856	2.847	8.130	26
27	3 303.8	.00030	.00011	.3501	9 436.5	2.856	2.849	8.137	27
28	4 460.1	.00022	.00008	.3501	12 740.3	2.857	2.851	8.143	28
29	6 021.1	.00017	.00006	.3501	17 200.4	2.857	2.852	8.148	29
30	8 128.5	.00012	.00004	.3500	23 221.6	2.857	2.853	8.152	30
31	10 973.5	.00009	.00003	.3500	31 350.1	2.857	2.854	8.154	31
32	14 814.3	.00007	.00002	.3500	42 323.7	2.857	2.855	8.157	32
33	19 999.3	.00005	.00002	.3500	57 137.9	2.857	2.855	8.158	33
34	26 999.0	.00004	.00001	.3500	77 137.2	2.857	2.856	8.159	34
35	36 448.7	.00003	.00001	.3500	104 136.3	2.857	2.856	8.160	35

40% **Compound Interest Factors** **40%**

	Single Payment		Uniform Payment Series				Arithmetic Gradient		
n	Compound Amount Factor Find *F* Given *P* *F/P*	Present Worth Factor Find *P* Given *F* *P/F*	Sinking Fund Factor Find *A* Given *F* *A/F*	Capital Recovery Factor Find *A* Given *P* *A/P*	Compound Amount Factor Find *F* Given *A* *F/A*	Present Worth Factor Find *P* Given *A* *P/A*	Gradient Uniform Series Find *A* Given *G* *A/G*	Gradient Present Worth Find *P* Given *G* *P/G*	n
1	1.400	.7143	1.0000	1.4000	1.000	0.714	0	0	**1**
2	1.960	.5102	.4167	.8167	2.400	1.224	0.417	0.510	**2**
3	2.744	.3644	.2294	.6294	4.360	1.589	0.780	1.239	**3**
4	3.842	.2603	.1408	.5408	7.104	1.849	1.092	2.020	**4**
5	5.378	.1859	.0914	.4914	10.946	2.035	1.358	2.764	**5**
6	7.530	.1328	.0613	.4613	16.324	2.168	1.581	3.428	**6**
7	10.541	.0949	.0419	.4419	23.853	2.263	1.766	3.997	**7**
8	14.758	.0678	.0291	.4291	34.395	2.331	1.919	4.471	**8**
9	20.661	.0484	.0203	.4203	49.153	2.379	2.042	4.858	**9**
10	28.925	.0346	.0143	.4143	69.814	2.414	2.142	5.170	**10**
11	40.496	.0247	.0101	.4101	98.739	2.438	2.221	5.417	**11**
12	56.694	.0176	.00718	.4072	139.235	2.456	2.285	5.611	**12**
13	79.371	.0126	.00510	.4051	195.929	2.469	2.334	5.762	**13**
14	111.120	.00900	.00363	.4036	275.300	2.478	2.373	5.879	**14**
15	155.568	.00643	.00259	.4026	386.420	2.484	2.403	5.969	**15**
16	217.795	.00459	.00185	.4018	541.988	2.489	2.426	6.038	**16**
17	304.913	.00328	.00132	.4013	759.783	2.492	2.444	6.090	**17**
18	426.879	.00234	.00094	.4009	1 064.7	2.494	2.458	6.130	**18**
19	597.630	.00167	.00067	.4007	1 419.6	2.496	2.468	6.160	**19**
20	836.682	.00120	.00048	.4005	2 089.2	2.497	2.476	6.183	**20**
21	1 171.4	.00085	.00034	.4003	2 925.9	2.498	2.482	6.200	**21**
22	1 639.9	.00061	.00024	.4002	4 097.2	2.498	2.487	6.213	**22**
23	2 295.9	.00044	.00017	.4002	5 737.1	2.499	2.490	6.222	**23**
24	3 214.2	.00031	.00012	.4001	8 033.0	2.499	2.493	6.229	**24**
25	4 499.9	.00022	.00009	.4001	11 247.2	2.499	2.494	6.235	**25**
26	6 299.8	.00016	.00006	.4001	15 747.1	2.500	2.496	6.239	**26**
27	8 819.8	.00011	.00005	.4000	22 046.9	2.500	2.497	6.242	**27**
28	12 347.7	.00008	.00003	.4000	30 866.7	2.500	2.498	6.244	**28**
29	17 286.7	.00006	.00002	.4000	43 214.3	2.500	2.498	6.245	**29**
30	24 201.4	.00004	.00002	.4000	60 501.0	2.500	2.499	6.247	**30**
31	33 882.0	.00003	.00001	.4000	84 702.5	2.500	2.499	6.248	**31**
32	47 434.8	.00002	.00001	.4000	118 584.4	2.500	2.499	6.248	**32**
33	66 408.7	.00002	.00001	.4000	166 019.2	2.500	2.500	6.249	**33**
34	92 972.1	.00001		.4000	232 427.9	2.500	2.500	6.249	**34**
35	130 161.0	.00001		.4000	325 400.0	2.500	2.500	6.249	**35**

45% **Compound Interest Factors** 45%

	Single Payment		Uniform Payment Series				Arithmetic Gradient		
	Compound Amount Factor	Present Worth Factor	Sinking Fund Factor	Capital Recovery Factor	Compound Amount Factor	Present Worth Factor	Gradient Uniform Series	Gradient Present Worth	
	Find *F* Given *P*	Find *P* Given *F*	Find *A* Given *F*	Find *A* Given *P*	Find *F* Given *A*	Find *P* Given *A*	Find *A* Given *G*	Find *P* Given *G*	
n	*F/P*	*P/F*	*A/F*	*A/P*	*F/A*	*P/A*	*A/G*	*P/G*	*n*
1	1.450	.6897	1.0000	1.4500	1.000	0.690	0	0	**1**
2	2.103	.4756	.4082	.8582	2.450	1.165	0.408	0.476	**2**
3	3.049	.3280	.2197	.6697	4.553	1.493	0.758	1.132	**3**
4	4.421	.2262	.1316	.5816	7.601	1.720	1.053	1.810	**4**
5	6.410	.1560	.0832	.5332	12.022	1.876	1.298	2.434	**5**
6	9.294	.1076	.0543	.5043	18.431	1.983	1.499	2.972	**6**
7	13.476	.0742	.0361	.4861	27.725	2.057	1.661	3.418	**7**
8	19.541	.0512	.0243	.4743	41.202	2.109	1.791	3.776	**8**
9	28.334	.0353	.0165	.4665	60.743	2.144	1.893	4.058	**9**
10	41.085	.0243	.0112	.4612	89.077	2.168	1.973	4.277	**10**
11	59.573	.0168	.00768	.4577	130.162	2.185	2.034	4.445	**11**
12	86.381	.0116	.00527	.4553	189.735	2.196	2.082	4.572	**12**
13	125.252	.00798	.00362	.4536	276.115	2.204	2.118	4.668	**13**
14	181.615	.00551	.00249	.4525	401.367	2.210	2.145	4.740	**14**
15	263.342	.00380	.00172	.4517	582.982	2.214	2.165	4.793	**15**
16	381.846	.00262	.00118	.4512	846.325	2.216	2.180	4.832	**16**
17	553.677	.00181	.00081	.4508	1 228.2	2.218	2.191	4.861	**17**
18	802.831	.00125	.00056	.4506	1 781.8	2.219	2.200	4.882	**18**
19	1 164.1	.00086	.00039	.4504	2 584.7	2.220	2.206	4.898	**19**
20	1 688.0	.00059	.00027	.4503	3 748.8	2.221	2.210	4.909	**20**
21	2 447.5	.00041	.00018	.4502	5 436.7	2.221	2.214	4.917	**21**
22	3 548.9	.00028	.00013	.4501	7 884.3	2.222	2.216	4.923	**22**
23	5 145.9	.00019	.00009	.4501	11 433.2	2.222	2.218	4.927	**23**
24	7 461.6	.00013	.00006	.4501	16 579.1	2.222	2.219	4.930	**24**
25	10 819.3	.00009	.00004	.4500	24 040.7	2.222	2.220	4.933	**25**
26	15 688.0	.00006	.00003	.4500	34 860.1	2.222	2.221	4.934	**26**
27	22 747.7	.00004	.00002	.4500	50 548.1	2.222	2.221	4.935	**27**
28	32 984.1	.00003	.00001	.4500	73 295.8	2.222	2.221	4.936	**28**
29	47 826.9	.00002	.00001	.4500	106 279.9	2.222	2.222	4.937	**29**
30	69 349.1	.00001	.00001	.4500	154 106.8	2.222	2.222	4.937	**30**
31	100 556.1	.00001		.4500	223 455.9	2.222	2.222	4.938	**31**
32	145 806.4	.00001		.4500	324 012.0	2.222	2.222	4.938	**32**
33	211 419.3			.4500	469 818.5	2.222	2.222	4.938	**33**
34	306 558.0			.4500	681 237.8	2.222	2.222	4.938	**34**
35	444 509.2			.4500	987 795.9	2.222	2.222	4.938	**35**

50%				Compound Interest Factors					50%
	Single Payment		Uniform Payment Series				Arithmetic Gradient		
	Compound Amount Factor	Present Worth Factor	Sinking Fund Factor	Capital Recovery Factor	Compound Amount Factor	Present Worth Factor	Gradient Uniform Series	Gradient Present Worth	
	Find *F* Given *P*	Find *P* Given *F*	Find *A* Given *F*	Find *A* Given *P*	Find *F* Given *A*	Find *P* Given *A*	Find *A* Given *G*	Find *P* Given *G*	
n	*F/P*	*P/F*	*A/F*	*A/P*	*F/A*	*P/A*	*A/G*	*P/G*	*n*
1	1.500	.6667	1.0000	1.5000	1.000	0.667	0	0	**1**
2	2.250	.4444	.4000	.9000	2.500	1.111	0.400	0.444	**2**
3	3.375	.2963	.2105	.7105	4.750	1.407	0.737	1.037	**3**
4	5.063	.1975	.1231	.6231	8.125	1.605	1.015	1.630	**4**
5	7.594	.1317	.0758	.5758	13.188	1.737	1.242	2.156	**5**
6	11.391	.0878	.0481	.5481	20.781	1.824	1.423	2.595	**6**
7	17.086	.0585	.0311	.5311	32.172	1.883	1.565	2.947	**7**
8	25.629	.0390	.0203	.5203	49.258	1.922	1.675	3.220	**8**
9	38.443	.0260	.0134	.5134	74.887	1.948	1.760	3.428	**9**
10	57.665	.0173	.00882	.5088	113.330	1.965	1.824	3.584	**10**
11	86.498	.0116	.00585	.5058	170.995	1.977	1.871	3.699	**11**
12	129.746	.00771	.00388	.5039	257.493	1.985	1.907	3.784	**12**
13	194.620	.00514	.00258	.5026	387.239	1.990	1.933	3.846	**13**
14	291.929	.00343	.00172	.5017	581.859	1.993	1.952	3.890	**14**
15	437.894	.00228	.00114	.5011	873.788	1.995	1.966	3.922	**15**
16	656.814	.00152	.00076	.5008	1 311.7	1.997	1.976	3.945	**16**
17	985.261	.00101	.00051	.5005	1 968.5	1.998	1.983	3.961	**17**
18	1 477.9	.00068	.00034	.5003	2 953.8	1.999	1.988	3.973	**18**
19	2 216.8	.00045	.00023	.5002	4 431.7	1.999	1.991	3.981	**19**
20	3 325.3	.00030	.00015	.5002	6 648.5	1.999	1.994	3.987	**20**
21	4 987.9	.00020	.00010	.5001	9 973.8	2.000	1.996	3.991	**21**
22	7 481.8	.00013	.00007	.5001	14 961.7	2.000	1.997	3.994	**22**
23	11 222.7	.00009	.00004	.5000	22 443.5	2.000	1.998	3.996	**23**
24	16 834.1	.00006	.00003	.5000	33 666.2	2.000	1.999	3.997	**24**
25	25 251.2	.00004	.00002	.5000	50 500.3	2.000	1.999	3.998	**25**
26	37 876.8	.00003	.00001	.5000	75 751.5	2.000	1.999	3.999	**26**
27	56 815.1	.00002	.00001	.5000	113 628.3	2.000	2.000	3.999	**27**
28	85 222.7	.00001	.00001	.5000	170 443.4	2.000	2.000	3.999	**28**
29	127 834.0	.00001		.5000	255 666.1	2.000	2.000	4.000	**29**
30	191 751.1	.00001		.5000	383 500.1	2.000	2.000	4.000	**30**
31	287 626.6			.5000	575 251.2	2.000	2.000	4.000	**31**
32	431 439.9			.5000	862 877.8	2.000	2.000	4.000	**32**

60% **Compound Interest Factors** 60%

	Single Payment		Uniform Payment Series				Arithmetic Gradient		
	Compound Amount Factor	Present Worth Factor	Sinking Fund Factor	Capital Recovery Factor	Compound Amount Factor	Present Worth Factor	Gradient Uniform Series	Gradient Present Worth	
n	Find F Given P F/P	Find P Given F P/F	Find A Given F A/F	Find A Given P A/P	Find F Given A F/A	Find P Given A P/A	Find A Given G A/G	Find P Given G P/G	n
1	1.600	.6250	1.0000	1.6000	1.000	0.625	0	0	**1**
2	2.560	.3906	.3846	.9846	2.600	1.016	0.385	0.391	**2**
3	4.096	.2441	.1938	.7938	5.160	1.260	0.698	0.879	**3**
4	6.554	.1526	.1080	.7080	9.256	1.412	0.946	1.337	**4**
5	10.486	.0954	.0633	.6633	15.810	1.508	1.140	1.718	**5**
6	16.777	.0596	.0380	.6380	26.295	1.567	1.286	2.016	**6**
7	26.844	.0373	.0232	.6232	43.073	1.605	1.396	2.240	**7**
8	42.950	.0233	.0143	.6143	69.916	1.628	1.476	2.403	**8**
9	68.719	.0146	.00886	.6089	112.866	1.642	1.534	2.519	**9**
10	109.951	.00909	.00551	.6055	181.585	1.652	1.575	2.601	**10**
11	175.922	.00568	.00343	.6034	291.536	1.657	1.604	2.658	**11**
12	281.475	.00355	.00214	.6021	467.458	1.661	1.624	2.697	**12**
13	450.360	.00222	.00134	.6013	748.933	1.663	1.638	2.724	**13**
14	720.576	.00139	.00083	.6008	1 199.3	1.664	1.647	2.742	**14**
15	1 152.9	.00087	.00052	.6005	1 919.9	1.665	1.654	2.754	**15**
16	1 844.7	.00054	.00033	.6003	3 072.8	1.666	1.658	2.762	**16**
17	2 951.5	.00034	.00020	.6002	4 917.5	1.666	1.661	2.767	**17**
18	4 722.4	.00021	.00013	.6001	7 868.9	1.666	1.663	2.771	**18**
19	7 555.8	.00013	.00008	.6011	12 591.3	1.666	1.664	2.773	**19**
20	12 089.3	.00008	.00005	.6000	20 147.1	1.667	1.665	2.775	**20**
21	19 342.8	.00005	.00003	.6000	32 236.3	1.667	1.666	2.776	**21**
22	30 948.5	.00003	.00002	.6000	51 579.2	1.667	1.666	2.777	**22**
23	49 517.6	.00002	.00001	.6000	82 527.6	1.667	1.666	2.777	**23**
24	79 228.1	.00001	.00001	.6000	132 045.2	1.667	1.666	2.777	**24**
25	126 765.0	.00001		.6000	211 273.4	1.667	1.666	2.777	**25**
26	202 824.0			.6000	338 038.4	1.667	1.667	2.778	**26**
27	324 518.4			.6000	540 862.4	1.667	1.667	2.778	**27**
28	519 229.5			.6000	865 380.9	1.667	1.667	2.778	**28**

Continuous Compounding—Single Payment Factors

rn	Compound Amount Factor e^{rn} Find F Given P F/P	Present Worth Factor e^{-rn} Find P Given F P/F	rn	Compound Amount Factor e^{rn} Find F Given P F/P	Present Worth Factor e^{-rn} Find P Given F P/F
.01	1.0101	.9900	**.51**	1.6653	.6005
.02	1.0202	.9802	**.52**	1.6820	.5945
.03	1.0305	.9704	**.53**	1.6989	.5886
.04	1.0408	.9608	**.54**	1.7160	.5827
.05	1.0513	.9512	**.55**	1.7333	.5769
.06	1.0618	.9418	**.56**	1.7507	.5712
.07	1.0725	.9324	**.57**	1.7683	.5655
.08	1.0833	.9231	**.58**	1.7860	.5599
.09	1.0942	.9139	**.59**	1.8040	.5543
.10	1.1052	.9048	**.60**	1.8221	.5488
.11	1.1163	.8958	**.61**	1.8404	.5434
.12	1.1275	.8869	**.62**	1.8589	.5379
.13	1.1388	.8781	**.63**	1.8776	.5326
.14	1.1503	.8694	**.64**	1.8965	.5273
.15	1.1618	.8607	**.65**	1.9155	.5220
.16	1.1735	.8521	**.66**	1.9348	.5169
.17	1.1853	.8437	**.67**	1.9542	.5117
.18	1.1972	.8353	**.68**	1.9739	.5066
.19	1.2092	.8270	**.69**	1.9937	.5016
.20	1.2214	.8187	**.70**	2.0138	.4966
.21	1.2337	.8106	**.71**	2.0340	.4916
.22	1.2461	.8025	**.72**	2.0544	.4868
.23	1.2586	.7945	**.73**	2.0751	.4819
.24	1.2712	.7866	**.74**	2.0959	.4771
.25	1.2840	.7788	**.75**	2.1170	.4724
.26	1.2969	.7711	**.76**	2.1383	.4677
.27	1.3100	.7634	**.77**	2.1598	.4630
.28	1.3231	.7558	**.78**	2.1815	.4584
.29	1.3364	.7483	**.79**	2.2034	.4538
.30	1.3499	.7408	**.80**	2.2255	.4493
.31	1.3634	.7334	**.81**	2.2479	.4449
.32	1.3771	.7261	**.82**	2.2705	.4404
.33	1.3910	.7189	**.83**	2.2933	.4360
.34	1.4049	.7118	**.84**	2.3164	.4317
.35	1.4191	.7047	**.85**	2.3396	.4274
.36	1.4333	.6977	**.86**	2.3632	.4232
.37	1.4477	.6907	**.87**	2.3869	.4190
.38	1.4623	.6839	**.88**	2.4109	.4148
.39	1.4770	.6771	**.89**	2.4351	.4107
.40	1.4918	.6703	**.90**	2.4596	.4066
.41	1.5068	.6637	**.91**	2.4843	.4025
.42	1.5220	.6570	**.92**	2.5093	.3985
.43	1.5373	.6505	**.93**	2.5345	.3946
.44	1.5527	.6440	**.94**	2.5600	.3906
.45	1.5683	.6376	**.95**	2.5857	.3867
.46	1.5841	.6313	**.96**	2.6117	.3829
.47	1.6000	.6250	**.97**	2.6379	.3791
.48	1.6161	.6188	**.98**	2.6645	.3753
.49	1.6323	.6126	**.99**	2.6912	.3716
.50	1.6487	.6065	**1.00**	2.7183	.3679